YOUR CAR CAN BE HAZARDOUS TO YOUR HEALTH

The Book Automakers and Politicians Prefer You Not Read

By

Basil M. RuDusky M.D.

6/10/03

ISBN: 1-4033-8307-3 (e-book)
ISBN: 1-4033-8308-1 (Paperback)

Library of Congress Control Number: 2002094955

This book is printed on acid free paper.

Printed in the United States of America
Bloomington, IN

1stBooks - rev. 10/15/02

Table of Contents

Dedication

This book is dedicated to car lovers everywhere, and to those in the auto industry who really care.

Basil M. RuDusky
MD, FACA, FACP, FCCP, FACC, FACFE
15 Public Square
Wilkes-Barre, PA. 18701

Disclaimer

The author makes no pretense as to the universality of the opinions given. The evaluations and opinions are based on the subjective evaluations of two individuals only, with 45 years of an exceptionally high degree of personal experience in car-purchasing, driving, road-testing and discussions with many individuals in the auto industry.

The author realizes that the senses and sensibilities of every person involve a great deal of variance. This book was written so as to be useful to the car-buying public as a guide to evaluate the status of their bodies and person. This would place them in a more selective and perspective relationship to the very costly and important car-buying experience so that this can be of value to their general health and well-being. It is therefore, not a recommendation to purchase or not purchase any particular make or brand of automobile, that is and must continue to be an individual choice.

The experience described herein should serve rather, as a guide for one to assess his/her own body and health in relationship to the comfort and health-safety measures which various cars can provide.

In addition, there are no implied concerns or experience as related to mechanical auto safety, crash worthiness, crash injury status, etc. This information can be very easily obtained from appropriate sources involved with automobile testing – the federal agencies and private insurance and engineering safety organizations which have data on almost every automobile, sport-utility vehicle and small truck sold in the United States.

Introduction

The combination of back and respiratory ailments accounts for billions of dollars spent on medical services, absenteeism from work, and payment of various sick benefits and other forms of lost and wasteful compensation.

It has been the author's experience and personal observation over many years that the automobile is frequently an unrecognized source or contributing factor in the causation of backache, fatigue and many other obvious and insidious problems and complaints involving the musculoskeletal and respiratory systems. In some cases, brain function may also be affected. It is not rare amongst owners of new automobiles to find many individuals complaining of "just not feeling right", "feeling strange", "feeling like something is wrong", or "I'm just not feeling myself." Many have sought medical advice, to no avail. Oftentimes, a diagnosis of back ailments of various sorts are unknowingly due to sitting in a car that has poorly designed and constructed seats. The feelings of "sickness" with vague and nondescript symptoms and complaints, also having gone undiagnosed, may be due to inhalation of noxious fumes from the new car smell, as well as from outside polluted air coming into the interior of the car. Only recently have some automobile manufacturers been paying more attention to the seats and air quality in their cars. The author has undertaken a freelance quest over the past fifteen years for improvement in automobile seats and interior air quality by way of personal conversations, letters, telephone calls and direct interviews with various representatives of the auto industry in the United States and abroad. The result has been the noticeable improvement in seat comfort and safety, the institution of air recirculation systems, and more recently, the placement of ventilation air filtration systems in many automobiles. There is still much work to be done on all counts, as "near perfection" has not yet been achieved, either in seats, ventilation systems, or the interior new car smell.

Part I

CAR SEATS

Basil M. RuDusky M.D.

CAR SEATS

Broadly speaking, the tires and suspension generally contribute to the overall comfort and safety of an automobile. The next most important commodity in an automobile is the seat. They provide not only comfort, but when properly designed and constructed, are also an important safety item. The problem is, you can change tires any time and as often as you wish, but with car seats, you are stuck for the duration.

Back pain, especially of the lower back (lumbosacral area) is the most prevalent medical complaint in modern society. It has been reported that 65% to 85% of the adult population suffers from some form of low back syndrome, variously described as some form of pain, aching, stiffness, muscle spasm and restricted mobility. When significantly severe, these symptoms interfere with one's daily activity, workability, ordinary walking and equally important,

3

sleep. The commonest conditions which predispose one to back symptoms of all types are osteoarthritis, degenerative disc disease, herniated disc disease, osteoarthritic spurs (osteophytes) and the commonly encountered lumbo-sacral sprain syndrome, with or without the aforementioned conditions. Often, a combination of these problems are present. Billions of dollars are spent or wasted each year by productive and non-productive individuals on various forms of medical, physiatric, chiropractic and surgical care. When one includes absenteeism from work, payments for various sick benefits and other forms of compensation, the figure becomes astronomical. It has been estimated that in the United States more than eleven billion dollars are paid annually for workers' compensation benefits for work related back pain! This is only a small portion of the total direct costs of the overall economic impact placed upon our society, which has been estimated to be upwards of seventy-five to one hundred billion dollars annually – just for back pain![1]

The automobile seat is one of the commonest and the most overlooked source involved in the production or aggravation of back complaints. Even though more attention is being given to present-day automobile seats, the majority of cars produced are fitted with seats that are inadequately designed and constructed for total musculoskeletal (anatomic) comfort and protection. Those companies whose brochures attest to the time and effort spent on the design and construction of their car seats are not only frequently deluding themselves, but most certainly the car-buying public. Even those who make claim to an "orthopedically correct seat" too often suffer from delusions of pseudo-grandeur.

After many years of extensive experience in testing car seats, it is the author's opinion that these companies have either wasted much time and money in the design and development of their car seats due to inadequate knowledge and personnel or had chosen consultants whose practical knowledge of the human body and its association with the influence of driving was woefully inadequate or downright inept. Having

5

owned fifty cars thus far, I can attest to the multiple inadequacies from almost every engineering and design standpoint. This does not include the many hundreds of other cars my wife and I have personally tested. Up to this very day, it is my opinion that the majority of car seats are poorly designed and constructed. A big part of the problem is obviously "penny-pinching" (cost cutting to save a dollar wherever possible, and cheat the buyer out of something he deserves). Today, the car buyer must spend either a small or large fortune to get a good car seat, and even at a huge cost, many are quite lacking in design and construction for maximum comfort (safety not withstanding). The commonest problems confronted are:

(1)   improper or inadequate padding and springing of the seat

(2)   seats not tuned to the type of suspension

(3)   lumbar supports not in the true and needed lumbar position

(4)   poorly   located   and   dangerously   hard headrests (as well as inadequately adjustable)

(5)   seats   not   contoured   to   anatomically accommodate the human spine

(6)   inadequate   electric   and   manual   seat adjustments

(7)   excessive and hard bolsters

(8)   bolsters that have hard edge rods so that the thighs are repeatedly injured on exit

(9)   lack of standard memory adjustments even in higher priced cars (i.e., 2002 Audi A-6)

(10) base of seat too small or not long enough for proper thigh support and comfort

(11) electrically   adjustable   seats   that   do   not automatically separate the bottom of the seat from the back of the seat simultaneously

Most of the lower and middle-priced automobiles have seats which leave much to be desired and confer seating upon their owners which can at best be considered barely adequate, and at worst, totally abominable. Some cars in these classes offer seats that are properly designed and shaped, but their construction is so pathetically cheap, that comfort is not humanly possible. It is not uncommon in todays' market to have to spend upwards of thirty to fifty thousand dollars and more in order to obtain a satisfactory car seat, and even in this high – priced range, many seats are lacking in any one or more of the above necessary attributes.

Many seats are excellent in design and give good body and leg support except for the upper back area. In these cases, proper support could have easily been achieved had their designers allowed for a slight (10 degree) forward angulation of the upper one-quarter of the seatback. This flaw is also a common cause of headrest malpositioning except when they are properly designed and constructed so as to be in a proper and comfortable position – a not too common event, but definitely improving. My complaint and conversations with the technical and marketing staffs of BMW North America was not without its rewards for the car – buying public. Those efforts led to the improvements in design and construction of what arguably to some, may be considered to be among the very best seats in the auto industry. From these efforts was born the BMW comfort seat, with its electrically adjustable upper back. It is the only seat I know of that can fit almost any type of back regardless of the occupant's problem. Presently, I would rate the BMW seats, no models excluded, as being the best in the industry, and even they can be improved by further attention to scientific and practical details.

Admittedly, one does not really need a lumbar support or multiple electrical adjustments and positioners to have a near perfect seat. I can only site one example to fit this comment – the seat of the discontinued Infiniti J-30 series – the seat which I ranked number one in the entire auto industry. It was a seat that I sorely (a pun) miss. It immediately gave its occupants that "wow" feeling – like a designer Italian suit or pair of gloves. That seat supported one's back and neck like no other since – all simply because of proper design and quality construction, including the selection of its leather, an important and often overlooked part of a good seat. Presently, the seats from BMW and Mercedes-Benz, rank in the "best" category. The seats in the Volvo models and the Corvette are among the very few that have headrests built into the design of the seats (no up-down or fore-aft movement) but because of their proper design and construction characteristics are amongst the most protective systems available. Unfortunately, the seat bottoms of both cars are too short for total comfort, and the headrests are still too far posterior in spite of excellent design and construction. This is especially noticeable in the Volvo S-40 as compared to the upscale series.

Porsche likewise has the integrated seat-headrest construction, but in my opinion, the headrest areas are too posterior for comfort.

Another seat that comes quickly to mind was that magnificent lounge and comfort chair in the 1991 Cadillac Eldorado – my wife's favorite seat – a masterpiece of design and quality construction. It was constructed with a slight forward tilt in its upper portion at our recommendation, but unfortunately, the managers at General Motors did not allow a proper headrest to be installed, because of the soon to be introduced 1992 models. These proved to be a typically American auto industry state of "retrogressive advancement." They lacked the necessary forward tilt of the upper seatback, had poorly designed and constructed headrests, and a lumbar support that never got down to the lumbar area. In addition, the selection of leather and the absence of proper pleating for the design of the seat also did not help matters regarding seat comfort or longevity against aging, wear and stretching.

Another seat of pleasing comfort was the specially designed seat for the short-lived Oldsmobile Trofeo – a truly remarkable piece of seat engineering and one to be really missed.

There are seats in the lower-priced cars that offer better upper back support and neck safety, but in general, they are too uncomfortable overall for continual daily driving and on long trips. In fact, the American automobile manufacturers for years had been putting better seats in their trucks than they did in their cars.

What is often not realized nor appreciated, is that you can have a well-designed and properly constructed seat, yet still suffer considerable back discomfort and trauma because of the lack of proper suspension to chassis to seat dynamics. In this regard, German manufacturers, particularly Mercedes-Benz, BMW and Audi are of unparalled superiority. Amongst the American manufacturers, Cadillac is still at the forefront with its introduction of the Seville STS and DeVille series. The old cliché, "you only get what you pay for", certainly holds true in this segment, but there are cars in the $40,000 plus price range that still cheat the consumer out of a good seat.

In order to achieve maximum body comfort, the seat must be designed and constructed so as to vibrate in harmony with the chassis and suspension, so that the shocks passing from suspension to chassis to seat are uniformly absorbed or dissipated in a harmonious and synchronous fashion by the entire system acting in unison. The most resilient suspensions over sharp, high frequency bounces are those of Mercedes-Benz, BMW, Cadillac, and Audi. The suspension of the

Volvo S-80 is too jarring for practical comfort, in fact, to the point where the well-constructed seats cannot fully do their job in shock absorption. This opinion is made for cars that have very good handling characteristics combined with comfortable ride qualities. These are the so-called, "sport sedans." There are many cars that have a comfortable ride, but they do not have the "road-holding and handling" qualities at higher speeds and in fast turns. The best of the sport sedans are a compromise between comfort and handling, some exceed in one characteristic over the other (ride vs. handling), and the buyer must rely on a proper and adequate road test of those cars in which he has interest, or disappointment and financial loss are often the end result of an inadequate road test.

What then, should the automobile buyers do and what should they look for when it comes to car seats?

In rating seats I use 7 codes:

1 = poor design (shape, pleating, overall size, and thigh support)

2 = poor construction (comfort, padding,

material)

3 = inadequate headrests (design, construction, adjustments, comfort)

4 = inadequate seat adjustments, driver

5 = inadequate seat adjustments, passenger

6 = inadequate lumbar support (position, type)

7 = inadequate rear seating (comfort, back and neck support)

Taking each category in order, begin with the overall design or shape of the seat. Does it fit the size of your body for your height, and weight? This should be the first determination one makes. Then "feel" the fit of the seat – including the bottom or base and the back. Does it support your lower, mid and upper back – either with or without adjustments? Next, often neglected, is the check for proper thigh support – an utmost necessity for all types of driving. Is the bottom of the seat long enough to avoid mid and lower thigh pressure? Concomitantly, does the seat bottom have adequate downward and forward tilt to ease pressure on the thighs? Make a determination as to whether you fit in the seat or on the seat. If your immediate feeling is that your sitting on the seat rather than in it – forget it, you'll never be comfortable. Finally, is the seat bottom and back wide enough for your bottom and back?

When evaluating construction, several important points are mandatory in this regard. Foremost is the material used. Seat padding is a science and cannot be gone into in great detail, for that is a lengthy manuscript in itself. The padding must have a combination of proper shape, depth and resiliency to support, distribute and alleviate pressure over long periods of time, yet retain these properties for the life of the car. Again, you must have the feeling that you are sitting in the seat, rather than on it. Of course, the seat foundation and springing are important, but this can, in part, be evaluated only by a road test over all conditions – driving at speed, in fast turns and on the bumpiest roads you can find. If you live in an area endowed by good politics and good roads, then you're out of luck. A trip to the state of Pennsylvania may be worthwhile, especially Northeastern Pennsylvania, where we have the best road-testing facilities in the world! More pointedly, the city of Wilkes-Barre ranks tops in this category.

Now go to the headrest. Most importantly, does it fit the back of your head and neck without having to tilt your head backwards for a distance no greater than one inch to two inches, otherwise forget it. Many good or acceptable car seats fail this simple test. I can name numerous cars, even expensive ones, where you must drop your head several inches rearward in order to reach the headrest – a true no, no – not merely for comfort, but importantly, for safety as well. You will never escape a whiplash injury with a headrest of this type, and as a passenger, you will never be comfortable. All you will be doing is trying to buy various pillows and towels to take up the slack and give yourself what the manufacturers cheated you out of – comfort and safety. On long trips it is the passenger who will enjoy the endowments given to proper headrest design, construction and position. Those antiquated, uncomfortably dangerous headrests which are known as the box-type, see-through headrests, still occasionally creep up in cars of present-day manufacture. They should be outlawed. They are generally made as hard as a rock, have inadequate adjustment, and are most uncomfortable. They can in fact, cause the very injuries they were supposed to prevent. These were prominent features of car seats in various Saab and Audi models of the distant past. My complaints to Audi were heeded, then Volkswagen took up the disgraceful project in their new Beetle. Fortunately, they changed their thinking for the 2002 model year and have replaced those disastrous headrests with the normal, standard type used by most manufacturers. These types of headrests should be banned from the industry, and I certainly never expected the American or European automakers to bring them back. The newly introduced Buick Rendezvous for reasons beyond my understanding have these worthless and totally ineffective and uncomfortable headrests, also placed in a few other General Motors cars. As a matter of fact, the entire seat in the new Rendezvous is a disaster area of design and construction in my opinion. Talk about regression! That's as bad as the old plastic decorations introduced on some of the first cars coming from Japan, or the vinyl roof-coverings used on numerous American cars in the past. Incidentally, other Buick models and the new Chevrolet Impala have also retrogressed to those hole – in – a box headrests. It is, after all, the passenger who is in the most restful and comfortable position. Unfortunately most manufacturers still relegate the passenger to second-class status by omitting memory buttons and often the same and necessary adjustments given to the driver.

There are no mechanical adjustments that can begin to compete with an electrical seat-adjusting system. That is fact, pure and simple. However, many electrical seat systems are inadequate or inferior. It takes more science and electronics than fore and aft adjustments, up or down, and recline. These are just the basics. In actuality, it is the initial inadequate design and construction of the seat that makes one require the greatest number of adjustments. Seat comfort and support to the lower back cannot generally be obtained unless the seat bottom and seat back are made to separate from each other in a highly calculated scientific pattern. This area of the seat, which is the junction of lower portion of the seat back and the back of the seat bottom is known as the "sink" of the seat. It is basic human anatomy and physics that require this automatic separation of the sink of the seat for proper adjustment needed for spine and hip alignment and comfort. When you raise or lower the seat, or move the seat fore and aft, and you get this automatic separation which pulls the bottom of the seat away from the back, you then will be more assured of lower back comfort, meaning less muscle spasm, decreased pressure and misalignment discomfort, and less hip strain.

When all of the above criteria for a good seat are met, you must then decide if you require a lumbar support, or if it is standard, what type of support it is, and will it do its job for you, as the occupant? Mechanical rod or bar lumbar supports are simply inadequate and uncomfortable. They offer the occupant little, if anything in the way of proper back support. Equally as inadequate are the inflatable back supports that are not properly positioned. It seems that the seat consultants for many auto manufacturers have difficulty in accepting the location of the human lumbar spine. All too common are inflatable supports that locate themselves in the area between the upper lumbar spine and lower thoracic spine – the area where support is least needed and of least value. This means that the inflatable support is too high (upwards) from the full lumbar spine and really will not support the area known as the "small" of your back, where most anatomic support is needed. In addition to location, the support must·be of adequate height, for one that is too narrow can create more problems than it was meant to solve. A minimum of four inches of properly contoured inflatable back support is generally required.

According to a recent study, back problems during long car trips were concluded to be due mainly to vibration. This was a report given by the orthopedic research department of the University of Vermont.

The report indicated that our spines vibrate at a natural frequency of 4.5 cycles per second, and when an environment is encountered in which these vibrations duplicate the natural frequency, the back is in trouble. I can assure you, there is much more to it than mere vibrations. All one has to do is sit in a car seat in a showroom for a long enough period of time, and you will begin to feel the discomfort of a poorly designed or constructed seat. Furthermore, the older you are, or the greater problem you have with your back, the sooner you will be able to differentiate between a good or a bad seat, even before your all important test drive.

Headrests are not only an important source of comfort, but are an extremely valuable mode of protection against many injuries. An excellent report by the insurance institute for highway safety has shown what I have been preaching for at least two decades.[2] The study concluded that few head restraints can be classified as good, and most were poor. It had also shown that head restraints with good geometry can reduce neck injuries, and since most cars did not have restraints with the necessary design and construction, they would not be protective of the occupants in rear-end collisions. In testing 164 cars,

21

117 restraints were rated as poor – certainly no shocking news to the knowledgeable and astute car buyer.

They rated Volvo automobiles as having the best restraints. I agree with the "restraints" or protective criteria, but I could not give Volvo the excellent headrest rating across the board simply because of the comfort deficiency imposed on driver and passenger. Even though the headrests are tilted or protrude anteriorly, only with the seat in the upright position do they come close to touching the back of one's head and neck, thus limiting the necessary comfort to the driver and passenger, mainly the latter. A slight anterior tilt of the upper seatback rather than their posterior tilt would have allowed for more comfort and best positioning of the fixed headrest, which is nicely shaped and padded.

The newly introduced Ford Thunderbird is a great example of a fair seat with, for all practical purpose, no headrest. I challenge you to reach it without bending your neck in half. After all, would you expect more in a 2003 model-year automobile when its big brothers and sisters (i.e. Lincoln Continental and Town car) suffer form the same malady? Yes, I would. But the manufacturers of many models place far greater emphasis on cup-holders than they do on your comfort and safety. Volvo and Mercedes have recently placed great emphasis on total seat safety. The most recent being the emphasis on "breakaway" seats whereby the total seat displaces rearward in a rear-end collision rather than just the usual collapse of the seat-back of a poorly and cheaply constructed seat. These systems are exemplified by Volvo's whiplash prevention system (WHIPS) and Mercedes-Benz front-seat strengthening system.

The CBS news program "60 minutes" presented an excellent program concerning the ludicrous status of federal safety regulations and manufacturers' neglect in affording the public access to safer car seats. The program was hosted by Ed Bradley and aired February 16, 1992, yet little has been done since that presentation of deficiencies which have been known since at least as far back as 1978.[3]

Basil M. RuDusky M.D.

Only a seat with a slight forward tilt of the upper back area can give adequate upper back and shoulder support. Very few seats in existence accomplish this task. My input into the recommended development of the BMW comfort seat with an adjustable upper seat back provides the best necessary positioning of the upper back for anatomic comfort. This also provides for better adjustment and comfort in positioning the headrest, which, even in this fine car, needs to be redesigned and redeveloped for a better combination of safety and comfort, as I soon expressed to the marketing and development people after my first road test.

Exaggerated seat bolsters are an unnecessary evil when it comes to daily – driven passenger cars. It is simply amazing how the manufacturers fall for the inane advice given by automotive journal car-testers for "more bolsters, more bolstering, inadequate bolsters, inadequate bolstering", just as they do for more speed, more torque, more road feel, and all those other frustrated boy-racer clichés which they use in writing their articles.

Surely, a certain amount of bolstering is nice to have for both the back and buttocks in order to keep you safely and comfortably in your driving position in order to prevent excess slipping and sliding of your body in fast turns and maneuvers. The amount these frustrated pseudo-racers desire is in excess of normal comfort for the average, non-racing, non-reckless driver and passenger.

I have tested and refused to purchase many fine automobiles simply because the side bolsters were "too tight", like an ill-fit suit, and the bottom bolsters were so high, and so hard, they injured your thigh and buttock muscles and your sciatic nerve on every exit from the car. Anyone who has suffered from sciatica knows the discomfort associated with that condition. Excess bolstering makes the entrance and exit process very difficult and uncomfortable. A few of the manufacturers have solved this problem by maintaining non-racing bolsters of minimal protrusion with excellent ordinary driving effect and enhanced comfort as well, by making the bolsters of a soft, collapsible material similar to the Volvo S-80 and

Audi A6 sedans. In short, the car seat must be fully supportive, but not obtrusive. You must be able to sit in the seat, not simply on it. It must give good support from the neck, to the upper back, mid-back, lumbosacral area and the often-neglected thigh support. It does the driver nor the passenger any good to have adequate back support then finish the driving day with sore thighs, a rather common occurrence in my lengthy and costly experience.

The only way to avoid thigh strain and its attendant pain is by having a seat with a long bottom (unfortunately an uncommon scenario), proper shape, very selective (costly) padding, superb electrical adjustments (especially one with seat bottom and back separators), and the best covering material with anatomically proper pleating and stitching. One of the very best in this regard was the 1991 Cadillac Eldorado which had the best combination of seat length, width, padding, pleating, stitching and leather I have ever owned.

Last, but not least, the final necessity to seat and body comfort and health is the most neglected area by the average car buyer. True, if you are young enough and supple, you may be able to tolerate almost anything except the extremes, but once you get beyond that point, everything counts, and every little bit helps. Leather, leather, and leather. Like every thing else in life, there is leather, and there is leather – just like silk, wool, cotton and plastics.   True, not everyone can afford the cars with fine leather, but any grade leather is better than none at all. Why do I emphasize leather? Simply because it is the final necessity for lower back comfort. It is impossible to protect your lower back, or to prevent additive problems to those which most of us already have without having a leather seat. It gives you the additional proper support, the icing on the cake so to speak. It also enables one to twist and turn with a sliding motion, and to shuffle, as well as change position in a satisfactory and comfortable manner. You are not "glued to the seat", and your lumbar spine requires less force to enter, exit, twist, turn and adjust your body, thus producing far less strain on your back, especially the all-important lumbosacral area. I have stated for decades, that one of the commonest causes of low-back strain is the automobile seat, next in line being the bed mattress, followed closely by the office or workplace chair.

What is most unfortunate is that the more expensive and even the most expensive cars often do not meet all of the above acceptable necessities and criteria for making the ideal car seat. There are very few things more disheartening in automobile ownership than paying upwards of fifty thousand dollars, only to find out a few hundred miles later, that your car seat doesn't fill the bill, and you just can't live with it. This is not an infrequent occurrence, I can assure you. Nevertheless, all cars are an expensive budget item, and at any price, all should meet the basic requirements necessary to promote health and safety.

Hopefully, by paying attention to the items discussed, you will be afforded a better opportunity of choosing the right car seat, and have a better chance of making a decision which you can comfortably live with. After all, for the non-boy-racer, the basic foundation of your car is, in the beginning and the end – the seat.

Part II

POWER STEERING

Basil M. RuDusky M.D.

POWER STEERING

Often times, the steering system of the car is ignored by the buyer. The uninitiated buyer thinks that just because the car has power steering, all will be well. Not so! Buyer beware, especially those with neck, shoulder, elbow and wrist problems These include bursitis, tendonitis, tenoynovitis, nerve plexus injuries in the neck and armpits, cervical disc syndrome, osteoarthritis, rheumatoid arthritis, carpal tunnel syndrome, and thoracic outlet syndrome.

There are varying degrees of power steering. Variable assist, with the least effort at low speeds in order to facilitate parking is a must for the driver with any of the above medical problems. Many of todays' automobiles claim variable-assist power steering, but are tuned to the young driver desiring the so-called sport-handling type of steering with a greater degree of "road feel."

Unfortunately, the manufacturers often heed the complaints of these individuals and the road-testers, which is exactly why I did not purchase a BMW 5 or 3 series model, the lovely Volvo coupe, or the Mercedes CLK coupe – one exit from the dealer's parking lot was enough for me, and my wife as well. The steering at low speeds required too much effort because of the engineering characteristics of the type of power steering in those cars. This was one reason why we purchased the Cadillac STS, Volvo S80, and Audi A6 as our daily drivers. They had the best combination of seating, steering and smelling, although all except the steering could still be improved in all three cars, even though they far exceeded most others in those categories.

When one has to expend a recognizable amount of effort in turning the steering wheel, the more likely it is to cause neck or arm strain. The muscles and nerves so utilized result in pain and strain due to stretching and tension placed upon the nerves, muscles and joints of the neck, shoulders, arms, elbows and wrists.

Aggravation of cervical (neck) disc syndromes, various neurologic states affecting the nerve bundles in the neck and armpits (axillae), as well as tendonitis, bursitis, arthritis and nerve entrapments in the elbows and wrists is more likely to occur. The results are various types of pain and numbness as well as restricted mobility of one's joints, depending on the association with any underlying conditions previously present, and the degree to which they have been aggravated by the steering effort. Therefore, it is very important for middle-aged and older persons, and all individuals suffering from any of the aforementioned problems, to buy a car that has minimal steering-effort at parking speeds and slow driving speeds around town. A simple test is to try to steer the car using only the thumb and two fingers of one hand. If this can be easily done, then the power-steering boost is appropriate for parking and turning at low speeds, and you will not suffer from the production and aggravation of the disabling conditions previously stated.

Basil M. RuDusky M.D.

Part III

VENTILATION SYSTEMS

Basil M. RuDusky M.D.

VENTILATION SYSTEMS

For nearly as long a period of time as was spent on studying car seats, I toiled on expressing my opinions, and getting the automobile manufacturers to pay attention to the antiquated ventilation systems in their cars. A personal, free-lance quest of many years is finally beginning to show results. I could not begin to calculate the time spent world-wide by way of letters, telephone conversations, visits to dealerships, visits to the manufacturers United States headquarters, attendance at automobile shows, personal discussions with vice presidents of marketing, development, and research and automotive executives in general. A few letters to the editor of national and international car magazines also stirred some interest within the car buying public as I shall later relate.

Worldwide concern over the increasing numbers and cost of respiratory diseases and sensitivities has

37

received sporadic attention amongst the public, the medical profession and far less so, with the bureaucracies of the industrialized nations. The costs relating to time taken from work, permanent disability, workmens' compensation, insurance benefits and medical care have amounted to incalculable billions of dollars per year. Significant strides have been made in reducing automobile and industrial pollution, but are still grossly inadequate, especially when it comes to diesel and industrial pollution.

In the year 1990, a noted health economist calculated that the cost to the United States economy for simple allergic rhinitis alone was 1.8 billion dollars. Allergic rhinitis is the medical nomenclature for hayfever symptoms affecting the nose and causing swelling of the nasal passages, excessive mucous production (watering of the nose), and difficulty breathing through the nose. Furthermore, approximately 22.4 million Americans or 9.3% of the population in 1988 suffered from this condition. He estimated at that time, 881 million dollars was spent on physician visits and 276 million dollars for medications.[4]   This is a remarkable figure for a problem that is still on the increase, but even more so, when one considers the fact that this is only a small part of the overall cost and economic loss when applied to the entire field of respiratory and pulmonary ailments which plague the animal world. None of the studies have estimated the costs and losses attributed to the rest of the animal kingdom, whether or not they get medical care.   Diseases such as pulmonary emphysema, pulmonary fibrosis, bronchial asthma, chronic bronchitis, enhanced respiratory sensitivity syndrome, bronchiectasis, pneumonia, lung cancer and various combinations leading to chronic obstructive lung disease are increasing and are more frequent in combination than is allergic rhinitis.

Multiply the previously given figures for rhinitis by at least fifty and you may begin to appreciate the cost of the aforementioned medical problems and this most assuredly would be an underestimate!

Recent newspaper editorials have stated that experts are at a loss to explain the increasing asthma epidemic that has hit inner cities especially hard. All one can say is that the ability of their brains to achieve an adequate degree of introspection, perception and deduction are badly lacking in power and substance. Asthma deaths alone have doubled over the past two decades in spite of the improvements in diagnosis and therapy. In Illinois alone, the asthma death rate reported by the CDC (Center for Disease Control) for people between the ages of 5 and 34 increased by 341 percent between the years 1979 and 1994.

Does it take much common sense and deductive reasoning to appreciate the simple fact that it is all due to environmental pollution? The combination of cigarette smoking, inhalation of second hand smoke, automotive exhaust, diesel fumes, industrial fallout, and the many chemical and biologic toxins in our every day environment are all playing havoc with our respiratory health, and are undoubtedly contributing to the production and progression of many other diseases as well. These include numerous types of cancers, heart ailments and birth defects.

A study editorialized in USA Today (Aug. 17, 2001) taken from the journal Science, noted that more people are presently being killed by pollution from cars, trucks and other sources than by traffic crashes and produce not only long term effects, but numerous short term effects as well. Many studies in fact, have shown that outdoor particulate air pollution is hazardous to human health.[5] Airborne particles are known to influence respiratory health and these influences vary according to particle size (coarse particles considered to be 2.5 to 10.0 microns in diameter and fine particles less than 2.5 microns in diameter). Placed in simple perspective, a particle less than 10 microns in diameter is less than one-seventh the width of a human hair. A micron is a unit of length equal to one-millionth (10 to the minus 6) of a meter. Studies on other effects of fine particles and children's respiratory health have shown that fine particles, particularly those containing sulfate components, were more significantly associated with asthma in children than coarse particles, which often produced symptoms such as cough.[6]

Air pollution with increased particulate levels, particularly those of nitrogen dioxide, pose a significant cardiovascular risk to patients with implanted cardioverter-defibrillators.[7] This study revealed that an increase of 26 parts per billion (yes folks, note that billion in the nitrogen dioxide content of ambient air) was associated with increased defibrillator shocks within 2 days of exposure. These were patients in whom a cardioverter-defibrillator unit was placed in their chest and hearts to administer an electric shock in order to terminate a life-threatening cardiac arrhythmia (a rapid and abnormal heart beat). Researchers at Yale University had shown that the air inside school buses is full of soot and toxic chemicals, 15 times higher in idling school buses than in the outdoor air. Almost 90% of school buses just happen to be diesel powered. In 1996, diesel exhaust contributed to more than one-fourth of the nearly 24 million tons of nitrogen oxide pollution in the United States! Diesel exhaust contains more than 40 different chemical compounds either in gaseous form or are attached to the fine particles, all believed to be cancer-producing toxins. The California Air Resources Board, which leads the nation in investigative and proposed legislation concerning air pollution, estimates that 71% of the airborne cancer risk in Southern California is due to the 2% of diesel vehicles. Furthermore it was estimated that students riding school buses are exposed to as much as 46 times the cancer risk as compared to that considered to be significant by the EPA. Government estimates revealed that new diesel school buses emit 51 times more toxic pollutants than a new natural gas school bus and even those buses certified in 2001 emit twice the nitrogen oxide and three to four times more particulate matter than does the natural gas engine. Apparently most diesel drivers and owners feel that having diesel engines allows them to idle endlessly while polluting the earth and poisoning all of the animal and plant life with irreparable damage. Delivery trucks, buses, ambulances and industrial equipment account for vast amounts of the earth's pollution with its resultant toxic effects on the entire animal and plant kingdoms. I have seen buses idle on our streets, often unattended for thirty to sixty minutes. What is even more disgusting, is the grossly apparent ignorance and stupidity (a dangerous combination of traits for any human being to possess) of the bus drivers who stand by these toxic waste dumps and (would you believe?) smoke cigarettes at the same time! Then by the time they are in their fourth or fifth decade of life (provided they are still alive), they will be applying for work-related disability or their widows will be seeking legal counsel for work-related compensation death benefits. I have seen idling ambulances outside of our hospitals polluting the emergency rooms and up to two levels of the hospital as they idle endlessly without the restrictions that should be placed upon them.

Legislation involving diesel engines and fuels has been dangerously slow in coming, and presently is a mere pittance of what should, could and needs to be done – thanks, as usual, to our politicians and bureaucrats, as well as the self-serving corporate mainstream. The natural resource defense council (NRDC) issued the following statements in a bulletin of April 25, 2002. (Its entire text can be obtained on-line at http://www.nrdc.org/air/transportation/ehc/chap4.asp).

While the EPA has regulated the content of diesel fuel and has attempted to require diesel manufacturers to lower emissions from those engines, it has taken a "hands-off" policy toward forcing engine manufacturers to produce engines that run on cleaner fuels. New heavy-duty engine standards adopted in 1997 for implementation in 2004 reflected the EPA's willingness to prolong the use of diesel engines. It further stated that even though these standards were 40% more stringent for nitrogen oxide than those of present engines they failed to require a reduction of the excessively harmful and also proven cancer producing particulates so prevalent in diesel exhaust. I have maintained for years, that diesel fumes are one of the most toxic substances in our air.

Have you had the common and unfortunate experience of driving behind any bus, especially a school bus? God forbid. Your day is ruined as well may be the next two days or more, especially if your car doesn't have an adequate recirculation air system, or better yet, one with an adequate air filtration system.

A diesel-engine bus or truck, operating on its routine daily schedule, emits enough soot alone to fill a one gallon container, and this is just soot, not counting the other poisons which are emitted. A study by the Harvard school for public health has shown a pronounced link between air pollution and heart disease deaths in the 6 American cities studied.

The EPA (Environmental Protection Agency) in 2000 has finally decided to declare diesel fumes a distinct health hazard (what most of us who are affected by them have known for years). They reported that diesel fumes are a "likely human carcinogen", meaning they are being implicated in the production of cancer. Even though diesel fumes are a disproportionately large source of pollution in our nation's air, those responsible, as is usually the case until its too late, deny and criticize these reports as being unfounded, with no basis in fact. How many times have the caring, concerned citizens and environmentalists been thwarted by those reactions? Too numerous to count, and in every walk of life for the living and working. The California Environmental

Protection Agency has rightly concluded that the particulate matter in diesel exhaust causes lung cancer. Their report noted that diesel exhaust particulates of 5 or more parts per million are associated with increased rates of lung cancer. Further facts indicate that even though diesel-powered trucks and buses contribute a small fraction of the road vehicles, they account for 26% of the nitrogen oxides and up to 70% of the soot in urban air. Many other toxins are also part of these highly poisonous diesel fumes. I cannot correlate the facts regarding diesel-powered trucks, buses and industrial equipment with the contaminates emitted by diesel-engined automobiles and the standards for same. I can state that several years ago, a representative of Mercedes-Benz N.A. informed me that the exhaust emissions of their diesel-powered cars were "clean", whatever that meant. If so, then heavy vehicle and truck standards should be the same. I can wholeheartedlly state that I have never known any American – made diesel vehicle that I could consider as having an acceptable smell coming from its exhaust.

TABLE A

A partial list of diesel fume toxic contaminants as listed by the California Environmental Protection Agency. (for complete report see NRDC bulletin, Chapter 2, "Human Health Impact").

acetaldehyde

acrolein

antimony compounds

arsenic

benzene

beryllium compounds

biphenyl phthalate

butadiene

cadmium

chlorine

chlorobenzene

chromium compounds

cobalt compounds

creosol

cyanide compounds

dioxins

dibenzofurens

ethyl benzene

formaldehyde

lead

manganese compounds

mercury compounds

methanol

methyl ethyl ketone

naphthalene

nickel

nitrobiphenyl

phenol

phosphorus

various polycyclic hydrocarbons

proprionaldehyde

selenium compounds

styrene

numerous xylene mixtures

California scientists have shown that the air inside traveling cars was dirtier than the outside air because of the pollutants caused by the vehicles they are following, and when the cars follow diesel vehicles, the levels of soot inside the car was 6 to 8 times greater. These researchers concluded that diesel fumes pose a higher cancer risk than all of the other air pollutants put together. I am in complete agreement with that conclusion. Numerous experimental studies have shown that diesel fumes cause mutations in chromosomes and damage DNA, therefore implicating a mechanism for their cancer-creating qualities. It was determined that diesel-exposed workers have a 30% increased risk of developing lung cancer. The efforts of the NRDC has brought to the forefront the experimental, ecologic and epidemiologic research substantiating diesel exhaust as a formidable and unequivocal health hazard. An American Cancer Society study found that people living in polluted urban areas had a 17% greater risk of mortality than those living in lesser polluted areas. In addition to the proven pulmonary and cardiac problems caused by the inhalation of diesel fumes, the various chemical compounds they contain are known to cause blood disorders such as leukemia, birth defects, reproductive problems, hormonal dysfunction and can affect the bodies natural immune defense mechanisms (See NRDC report, Chapter 2, 4/25/02). In chapter 3 of the NRDC Diesel Report it was stated that legislated use of the newer diesel fuel formulations only slightly reduces the emission of nitrogen oxides and particulates, and worse yet, that greater than 95% of the particle emissions were less than 1 micron in size. In addition, their investigation noted that the quantity of many of the other chemical toxins remained essentially unchanged. It has been clearly and well substantiated that increased air pollution can and does cause increased morbidity and mortality.[8] The larger the particle (>5 microns) the more likely it is to become deposited in the upper respiratory tract (the larger airways) and the smaller the particle, the deeper into the lung it goes. Particles of less than 2.5 microns are more likely to be deposited in the small airways (bronchioles) and lung substance (alveoli).

If one has inherent lung damage or disease from whatever cause, such as cigarette smoking, the irritation and damage to the lungs is greater than would ordinarily be expected. Particulates in the air include ozone, carbon monoxide, sulfur oxides, nitrogen oxides, hydrocarbons, carbon dioxide, methane, various dusts and pollens and numerous other chemical pollutants. Exposure to low levels of carbon monoxide is usually tolerated by healthy individuals, but can be problematic to individuals who suffer from anemia, coronary heart disease or lung disease. Sulfur dioxide produced by various combustion processes is one of the components of acid rain (as sulfuric acid), the others being nitric acid and hydrochloric acid. If acid rain damages the paint on automobiles, just imagine what it does to your lungs. Ozone in the upper atmosphere is beneficial in that it protects the animal and plant kingdoms from the harmful effects of excess ultraviolet light. In the lower atmosphere, its inhalation causes cough, chest pain and shortness of breath. Numerous volatile organic compounds have been either implicated or suspected in producing harmful effects on the human body. These are often the cause of numerous complaints produced by the "sick-building syndrome", as well as, I might add, the "sick car syndrome", or new car smell, which will be discussed in some detail subsequently.

Previous and recent studies have shown a distinct relationship between excess cardiovascular morbidity (sickness) and mortality (death) after exposure to particulate air pollution.[9] In fact, progression of atherosclerosis (blockage of the arteries by clot and fat), as well as an increased vulnerability of these abnormal plaques within the arteries to cracking or rupture, producing angina (chest pain from the heart), cardiac arrhythmias (irregularity of the heart beat) and myocardial infarction (heart attack) have been shown to occur in experimental, clinical and epidemiologic studies.

Another recent study concluded that long-term exposure to combustion-related fine particulate air pollution is an important environmental risk factor in the causation of increased cardiac, pulmonary and lung cancer mortality.[10]

The above has given only a brief and limited view of air pollution and some of its effects on our health. The automobile has become the quintessential force in the modern, industrialized society. It is not only a creator of harm, but can be at least partially transformed into a health benefit for its occupants.

Over nearly a two-decade period of time, the author has struggled with the automobile manufacturers throughout the world to use recirculation ventilation systems and air purification systems in their automobiles. The free-lance, public-service and personal self-serving interests of the author have shown some satisfying results, but much more can be done and needs to be done in the further development and introduction of air filtration and ventilation systems that allow us to breathe the best possible quality of clean air coming into the interior of our cars. As I stated to some of the manufacturers, "you may as well do it now, before it becomes a federal mandate", which of course, it should become. But, as we all know politicians and politics in the United States, the sole purpose of their existence is the next election – to hell with the real people who work themselves to the bone and pay all the taxes and obey all the laws. Their democratic-socialistic attitude is hurrah for the minority causes, immigrants, welfare recipients, criminals and every other country and peoples in the world other than our own and least of all those that are the necessary productive core of our society. They almost never (or is it actually, never?) pass a law that costs the taxpayer nothing – a law that would benefit society at no cost, just profit. Perhaps it's because they can't scheme, coerce, wheel and deal and waste with these types of laws. And of course, the hard-working people can't afford to demonstrate, for they are too busy keeping the lawmakers in all their finery, with all the perks and the continual salary increases they allot themselves. Typical examples of their inaction over the decades has been the absence of non-smoking laws with restrictions in all public places, including streets and transportation, anti-automobile idling laws, and anti-pollution measures for small-engine manufacturers, lack of adequate diesel regulation not withstanding. Laws embodying these three measures alone would save billions upon billions of dollars in preventing pollution and contamination of our entire ecologic system and environment, as well as billions of dollars in present and future health costs, disability, workmens' compensation, social security, and more. But that would be all too simple and cheap to initiate, and maybe would cost someone a few votes in the offing, so why bother, especially since there is no large organization or so-called minority group that cares about the people as a whole, just one group or coalition after another with self-serving interests and future goals based on personal prejudices and disguised as pseudo-constitutional freedoms and rights. Another example is the lack of regulation regarding the manufacture and use of cheap, toxic interior car materials. Do you think a politician can put that degree of advanced thinking into his daily or yearly schedule? God forbid! He wouldn't then be spending and wasting our hard-earned, and becoming more difficult to come by, tax dollars.

What is equally amazing, is just how reluctant the automobile manufacturers are in taking free, valuable advice, yet they pay millions of dollars in salaries and contracts to people who either don't know or simply are not doing their jobs as they should be. As previously noted, they still seem unknowledgeable about, or don't care enough to manufacture good car seats. Finally, virtually every car manufacturer now offers recirculation air systems in their vehicles. (See figure one). These simple systems which restrict or prevent exterior air from entering the vehicle are a very high priority comfort and health measure which had been neglected by the auto industry for decades. It took the author ten years to get the necessary individuals at Cadillac to install this simple necessity in their cars. When they did, it was not as beneficial as those systems of other manufacturers simply because of their free-flow or flow-through antiquated ventilation system that allows an excess of outside air into the vehicle even when the ventilation system is turned off. Still, better than nothing at all, as it had been for years. During my quest for the recirc air systems, the various comments from the auto industry giants were sickeningly ludicrous. "Our cars heat and cool faster", which was actually stupid, because in the owners manuals one can invariably find the statement that one can utilize the recirc button to assist in cooling or warming the interior of the car more quickly. Other comments were equally disconcerting. "The average person wouldn't know how to use it." "It fogs up the interior windows and would exasperate the driver." "It would cost a few dollars more per car", and some such comments which I have since forgotten. But, we finally have it, and it is here to stay, thankfully. The best system will shut off almost all of the incoming air, and the very best systems work in conjunction with the air conditioning system. Some systems are on timers, such as after fifteen minutes, will turn off, others are manual, with operator selection mandating the times on and off. I prefer the latter, because my brain still works, and I have found it an annoyance to be fiddling with the dash buttons every ten or fifteen minutes. I prefer lowering the windows every now and then to allow outside air to enter when I'm driving through a relatively clean area or to demist the windows in conjunction with the air conditioner when necessary. The most sophisticated systems work in conjunction with a ventilation filtration system that has automatic sensors which detect a given level of pollutants (toxic

Fig. One

Typical Example of Recirculation Air System

Button with semicircular arrow indicates ventilation system in recirculation mode. Open arrow indicates outside air.

gasses) coming into the car so that the recirc system comes on automatically and after a certain period of time, turns itself off. These can be found in cars by BMW, Mercedes and Audi.

After a successful quest in pursuing the worldwide auto industry greats to place recirc air systems into their cars, a ten-year struggle ("mein kampf") has been still taking place in attempting to get them to place ventilation filters in their cars. I have been continuously making calls, leaving messages, writing notes and talking to representatives of nearly every auto manufacturer. The recirc air system is certainly far better than nothing, as has been the case for so many years, but an automobile without a ventilation filter is now antiquated when it comes to air conditioning and ventilation. By my best recollection Saab was the first manufacturer to place such filters in their ventilation system many years before any other manufacturer. I then continued my greater than a decade quest for cleaner interior air in the automobile by prodding and requesting the manufacturers to use ventilation filters in their cars.

Following initial discussions with the authorities at BMW, it took them only 2 model years to comply, and with Mercedes-Benz, 3 years. Both BMW and Mercedes are to be highly complimented on adapting these systems so quickly.

After having a phone discussion with an engineer at the Corvette plant in Bowling Green, Kentucky, I harshly relayed the message that a Corvette is nothing but a fast street sweeper. I informed them that it sucks up everything into its air vents and poisons the occupants with same. Within 2 model years, a recirc air system was placed in the Corvette. I proceeded to test its effectiveness by taking a new Vette off the dealer's lot, went on a small hill which contained a signal light, pushed the recirc button and went behind an old Greyhound bus, spewing black poison from its exhaust, followed it for a few hundred yards, and beheld the fact that nearly all fumes were prevented from entering the car. They still haven't gotten around to placing air filters in their cars however, but in time, it will surely come – especially if more buyers would do just a little of what I had begun – complain and offer good advice.

While my battles for the war against polluted air having free ingress into the car interiors continued relentlessly, a fine gentleman by the name of George Morehouse, a consulting engineer, formed a company called Interior Air Quality, Inc.

After 5 years of research and development his company developed after market, add-on smog filters that could be ordered for some of the more popular model automobiles. These were designed to be placed into the air intake system of the cars, and they were very successful in removing large quantities of toxic gasses, dusts and pollens from the incoming air and were effective for 4000 to 12,000 miles of driving, depending on the degree of pollution encountered. I received a letter from George dated Feb. 10, 1993, indicating that he had ceased his manufacture of these filters because the auto industry is beginning to conform to the consumer needs and requests. The only problem was, that they have been lax and neglectful in doing so. His company, based in Colorado, used activated coconut shell carbon as the medium for fume removal. He developed his system because of personal chemical sensitivities to exhaust fumes and many other environmental pollutants. His filters contained the necessary activated carbon (3 units to be exact, making them more efficient than present manufacturers' units), and the presently utilized synthetic fiber filter to reduce pollen and dust. His advertisement and marketing brochure was simple, factual and straight to the point. It asked the following questions:

Do you have chemical sensitivities?

Are you sensitive to pollen and dust?

Do your eyes, nose and throat become irritated in heavy traffic?

Do you have breathing problems?

Do you suffer from environmental illness?

The brochure further stated: "The smog filter is especially effective in reducing airborne pollutants, including":

auto emissions

diesel fumes

ozone

nitrogen compounds

hydrogen sulfide

gasoline fumes

mercaptans

pollen and dust

smog

exhaust odors

Also noting that "activated carbon is not effective in removing carbon monoxide, as only a small percentage would be removed."

I knew that progress in the development and placement of vent filters would be generally delayed, as most manufacturers, especially those in Detroit, are very reluctant in taking free advice, regardless of how beneficial or accurate it may be. After all, 50 cars and 35 years of purchasing and testing, gives me at least a little knowledge and insight into the minds of the automotive industry.

It was then that I decided to fulfill my own requirements for cleaner interior automobile air in my personal cars. I began utilizing various electrostatic paper filters and carbon-impregnated filters used for various commercial air cleaning systems, rotating ceiling fans, in-house air cleaners, etc. I devised several systems of varying degrees of effectiveness and after cutting them to fit the grates of the air intake system near the cowl of each car (under the hood) where the air enters the vehicle, I would use black duct tape to hold the filters in position, and change them annually, or more often. A time-consuming task, running 4 cars per year, but certainly better than nothing, but not quite as effective as the present systems of BMW, Saab, Mercedes and Audi.

Presently (2002) many of the worlds' auto manufacturers have cars that can be purchased with factory or dealer-installed dust and pollen filters, known as microfilters. Again better than nothing, but still antiquated and far behind the leaders of the industry, all of whom have activated charcoal or carbonite systems to remove various incoming fumes from the highly polluted air. I was pleased to hear that in discussion with some representatives from Saab at the 2002 New York Auto Show, that presently, all of their vehicles contain activated charcoal systems for fume extraction. Although I cannot verify same, a representative for Volkswagen made the same statement.

BMW, Mercedes and Audi, all have activated charcoal systems in their cars except for some of the smaller sport models, which still have only true interior filtration systems, meaning dust and pollen filters.

Audi was the last of the big three German manufacturers to submit to intelligent reasoning. Porsche is also fitting their cars with air filtration systems. The biggest success story was during a recent trip to Italy I was told that the Alfa Romeo 166 series (a magnificent work of automotive art and sculpture) has an activated charcoal ventilation filter, without any prodding from me, I might add. Now that is a really remarkable event – for the Italian manufacturers to keep up with the others when it comes to comfort and convenience logistics, and actually is a step ahead of many of the American and Japanese manufacturers!

The microfilters presumably remove non-fume particles from 5 to 2 microns in diameter, the 2 micron systems being much more effective and giving far greater air quality inside the car. The need to purchase and change these systems on a 2 year or 20,000 mile cycle is well worth the price, and is the best thing any automobile buyer can do to preserve a portion of his health. Cadillac, usually one of the last to take advice, but not completely neglectful, placed a carbonite filter in its STS model, but after personal inspection of the filters, the amount of carbonite is inadequate, the microfilters are too porous, and when combined with the less effective recirc air system, fails to give the interior air quality of Saab, BMW, Mercedes and Audi. Additionally, 58,000 miles of driving an STS, 3 maintenance filter changes plus annual replacement of my external filters allow me to make that statement with utter impunity.

Even worse, the newly introduced Cadillac CTS series has only a simple microfilter, and yet this is supposed to be a car to compete with the likes of BMW, Mercedes and Audi – well, perhaps in cornering capability, but certainly, not in air quality, interior quality and aesthetics.

Recent data has linked particulate air pollution containing fine particles less than 2.5 microns in diameter to acute cardiovascular events occurring within 24 hrs of exposure. Studies throughout major cities of the world have shown an increased incidence of cardiac death-related events on the day following high levels of particulate air pollution as well as increased hospital admissions due to cardiac related causes, including acute myocardial infarction (heart attacks) and cardiac arrhythmias (irregularity of the heart beat and rhythm). The increased risk was stated as being approximately 62% for each 20 microgram per cubic meter increase in fine particles less than 2.5 microns in diameter.[11] The latest research in Germany has shown clearly that traffic related air pollution is an independent risk factor for the development of atopic eczema, a very annoying and problematic, allergic-type skin disease. It was believed that higher concentrations of small-diameter diesel exhaust particles, nitrous oxide, and volatile organic compounds such as benzene and toluene were probably the culprits. Still, the world's auto manufacturers continue to turn their backs and brains to the simple facts, merely for a few more dollars of profit per car. Equally so, the politicians do the same, for it would cost them nothing to implement proper legislation, but it would not be an attractive vote getting agenda that would stimulate the unintellectual public, "minorities", and numerous other groups of demonstrators who demonstrate for various causes of self-supporting vagaries of which they know very little and care even less about the subsequent consequences which may result from their actions.

The plain and simple electrostatic microfilters are adequate for removing large dust particles and various pollens, but do very little to filter out the fine particulate matter of various fumes, gases and odors found in the environs of our streets and highways. A modified HEPA (high-efficiency particulate air) filter would be better, but for automobile systems, its high resistance would require a technologic redesign of the filters. Activated charcoal, activated carbon, carbonite and better still, zeolite filters are necessary for adequate fume, gas and odor removal.

High quality home filters can remove up to 99-plus % of all pollen, dust, smoke, and particles as small as 0.3 micron.

Ford and GM have simple microfilters in some of their cars and of this writing, Chrysler has none in any model or series except, I am told, in one of their vans.

In summary, the industrial world has created an exceptionally harmful environment to all species of life. Automobile fumes now "stick to the ground", oftentimes even in rain, and take a longer time to dissipate than they did prior to the catalytic converters and newer gasoline formulas. The fumes are more irritating to the lungs, and toxic to the entire human body. Occupants of automobiles without air recirculation and filtration systems can detect the harmful fumes coming into the cars at least a half-mile or more away from the source. Diesel fumes are one of the most irritating, sensitizing, debilitating and dangerous fumes one can inhale – poison, plain and simple – simply deadly. Federal legislation has lagged seriously behind the needs of humanity, and the present unlegislated, poison-producing diesel trucks and buses will be on our roads for decades to come, for just like roaches, they're hard to kill. If they cannot be legislated to become significantly cleaner, they should be outlawed as a human hazard. With the advent of the catalytic converter, the increased presence of inadequately legislated and manufactured diesel vehicles, and the ever-increasing industrial pollution, physicians have seen more cases of acute, subacute and chronic pulmonary problems due to the irritation and sensitivity caused by these noxious pollutants over the past 10 to 15 years than during the previous 2 decades. The increasing rates of lung cancer, when added to the above, as well as the recently described cardiovascular complications, chronic obstructive pulmonary disease, bronchial asthma and various chemical sensitivities, including the well-documented "sick building syndromes", are costing the American taxpayer and associated federal and private insurance agencies multi-billions of dollars annually. Why? Simply because of ignorance and greed, both private and corporate, as well as governmental. The costs of enactment are infinitesimal when compared to the cost of alleviating the damages produced by them.

Part IV

THE NEW CAR SMELL

Basil M. RuDusky M.D.

THE NEW CAR SMELL

Unrealized by the majority of the car-buying public, that new car smell is also hazardous to your health – in some cases, it is very noxious and can be toxic. These noxious fumes emanate from all the interior materials, as well as the plastics in the ventilation system. The chemicals used in the dyes, adhesives, plastics, glues, in leather processing and tinting, carpeting and cloth seating are highly irritating and sensitizing substances. In some cases, they have not only provoked illness, but some individuals have become so sensitized to these chemical vapors that they now must seek older, used cars for purchase in order to avoid becoming ill – that is, if the particular car is able to give up its toxins in a year or more. This is not always possible, for some cars will give off the poisonous, sweet-smelling, sickening stench for years.

I have had reports of cars being taken back to the dealers, cars traded in only after a few weeks or months of ownership – all at considerable financial loss to the owners and occasionally to the dealer.

I had previously discussed the matter with various manufacturers' representatives including BMW, Infiniti, Lexus and Acura. With the introduction of the 1992 Cadillac Eldorado and Seville series, the "nuance" leather smell became quite annoying to some owners, including yours truly, who traded the 1992 Eldorado and Seville after only a few months ownership because of the lack of a recirculation air system, the sickening new car smell, and inadequately designed seats. I was able to find a black 1999 Seville STS with black leather that fortunately did not smell very much after a few weeks of ownership, but as stated previously, has a long way to go when it comes to a superiorly engineered ventilation system. The Infiniti representatives, as well as those from BMW, Mercedes and Audi stated that they were working on the problem (1993). They stated that they were trying to control the problem by proper selection of materials and the way in which those materials are manufactured. The people at Mercedes felt that they had substantially solved the problem, and I will admit that their cars were the least smelly of the lot at that time, with the exception of those with the cheap plastic seats which should never have been placed in their "lower-line" of cars. Presently, Porsche and Audi were the first to select "non-irritating" and stringently processed materials for the interior of their cars, and were soon followed by BMW, with Mercedes constantly improving their search for decent smelling automobile interiors. It is very easy to seek out and select a quality car interior if one chooses to do so. It is just a matter of being aware and paying attention to one's God-given senses – except if you're a smoker. They can't smell much of anything except severe stink, and can taste mainly salt and sugar containing foods and vinegar.

In case you haven't had the personal experience or opportunity to observe for yourself, the cars from Asia are the worst examples in the smell or stink category.

Other than Acura, one of the most sickening cars I have had the displeasure to sit in was the previous "low-Priced", BMW 318i with plastic seats. On the opposite end of the spectrum were clean and decent smelling and almost non-odorous cars such as the 1991 Cadillac Eldorado, Lincoln Mark VIII and the Mercedes 380 coupe. Other cars without harmful odors which had no effect on my senses whatever were the Porsche 928 series, the old Chrysler 300 series, the Volvo S-80 and various Audi models. The expensive Italian sport cars such as Ferrari and Lamborghini cars are also "nice-smelling" cars, with the odor being that of "pure", high-quality leather and interiors.

Some owners have told me that their Acuras and Hondas still smell badly after several years, and that they feel that their cars will "smell forever." Probably very likely. I have a 1977 custom-ordered Avanti II that still smells from chemical vapors within the interior of the car. It has been in storage, windows down, since that time and 25 years later still has that obnoxious new car smell. A discussion with an individual who has a similar awareness of car smells as

I do, jokingly told me that the Japanese are trying to "get even" by shipping us poisonous cars. The almost unbelievable Acura Legend story will be forthcoming.

The automotive industry as a whole, is quite aware of this problem, but as usual has been reluctant to admit to its existence or its seriousness. Their usual response is either indifference, or as our government bureaucracies so frequently state – "there is insufficient evidence."

I have in the past, contacted the U.S. Department of transportation, the National Highway traffic Safety Administration, the Center for Auto Safety (which was then a division of Ralph Nader's Consumer Protection Agency, and several independent research companies. It appeared at that time (1991) that only the Center for Auto Safety and several other independent institutions were concerned about the significance of the "new car smell" as being hazardous to one's health. Unfortunately they did not have the funds or support that would have enabled them to fully and properly study the problem in greater depth. I had done a personal, subjective survey of automobile showrooms, sitting in the new cars, taking them for test drives, and even purchasing a few for longer-term evaluation.

Year after year, I reached the same conclusion – the Asian-built cars were the most noxious and sickening to sit in and live with. Some cars will dissipate most of their toxic odors within one year, but many of the Asian-built cars continue to emit noxious odors for a much longer period of time, even on the used car lots.

Think seriously about this next simple question of fact. Why have an effective recirc air system in a car whose interiors are flooded with toxic fumes from within? Using the recirc air system in these cars will only cause the occupants greater harm by compounding the persistence of increased quantities of these harmful fumes because of the absence of incoming fresh air which helps in their dilution.

I was pleased to note that in the April 2001 edition of ROUNDEL, the official journal of the BMW Car Club of America, and I might add one of the best car club magazines published, was an excerpted article titled, "But Inside New Cars, Pollution Continues." It made me smile and say to myself, "what have I been saying for the past 15 years"? The study amazingly, came from a (would you believe) Japanese research group and concluded that the new car smell could make you sick. They measured "out-gassing" from the test cars interior leather, plastic, foam, vinyl and glues, and found 114 harmful volatile compounds coming from the car's interiors, including, formaldehyde. The news gets worse, folks! The new car cabin was 45 times worse than the World Health Organization's standards for indoor air quality! They further noted that the emission decreased over time, with the formaldehyde being mostly dissipated after 5 to 6 months, but the noxious emissions again rose during the warmer weather.

We are finally beginning to get somewhere regarding these harmful new car interior pollutants. Published in the May, 2002 issue of POPULAR SCIENCE, was an excellent article titled, "Gee Your Car Smells Terrific! You feel OK?" It was subtitled, "A scientist stalks the scent of a new machine." It stated that the real new-car aroma could be hazardous to your health. No kidding, has my message of concern finally reached places where something will be done about it? It stated symptoms caused by these smells were headaches, irritated eyes, and drowsiness. I will add many more in the upcoming section. It further noted that the researchers in the Renault Technological Center in Paris were quoted as saying that almost every car manufacturer in the world is concerned about new-car smell. Well, how many more years and decades must the car buying public suffer and possibly die until they do what should have been done many, many years ago? Many of these chemicals are toxic to the nervous system, the immune system and are cancer-causing toxins. The article centered about a special sensing mask developed in Spain, that recognizes some of the many odors coming from various portions of a new car. These smells consist of more than 200 left over organic solvents among which were ethylacetate, toluene and xylene, coming from more than 100 parts of the car. Although the present BMW automobiles have predominantly a clean, non-toxic smell which dissipates over a short period of time, it took my 1993 BMW 740i several years to "come clean", and the 1991 BMW 850i about 9 to 10 years to clean itself out. But, in contrast to the poisonous Acura story, neither I nor my wife became ill in the BMW cars.

In 1991, the Center for Auto Safety, housed in Washington D.C., sent me a compilation of some very interesting data and reports concerning carcinogenic (meaning cancer-causing) nitrosamines and plastics in automobiles. A letter from Clarence Ditlow to the then administrator of the National Highway and Traffic Safety Administration, Joan Claybrook, stated that researchers found that a human being sitting in the average new car for just 100 minutes would inhale 1 microgram of nitrosaimnes. The EPA concluded that as a family of carcinogens (cancer causing substances), the nitrosamines had no equal. This letter was written in 1979! Little has been done since that informative request was written – at least by our government, but some manufacturers, especially those from Germany, have taken great strides in this direction. The nitrosamines in high concentration can come from tires (a good reason to avoid those vehicles with in-vehicle tires such as exemplified by the harmful interior of the 1996 Jeep Grand Cherokee), other rubber products, and vinyl fabrics, in addition to many other parts of the automobile interior. The research report concluded that the nitrosamines were known to cause liver, lung and kidney cancers in experimental animals and that the concentrations absorbed by new car passengers is equal to moderate beer-drinking and higher than that of eating cooked bacon, one of the highest commonly encountered sources of nitrosamines. The copy of this excellent report which I received revealed that there was indeed a wide variation of these new car toxins amongst the manufacturers, but unfortunately, I was unable to decipher the codes indicating which models of which manufacturers were worse. The letter from Joan Claybrook to Mr. Ditlow was dated Oct. 1979. It was a typically bureaucratic response sharing their concerns and relegating the responsibility amongst the manufacturers and other agencies, noting the EPA (Environmental Protection Agency) has the charge of such problems. All I can say is that if they worked on this problem the way I see them addressing industrial pollution, God help us all!

A response to Claybrook and Ditlow by the two deputy assistants of the EPA also shared their concerns with the above, and in a letter of March 1980, they stated that they would assess the need for regulatory action concerning this problem, citing a very nicely written program to assess the problem.

Interestingly, a report published in Automotive News, Oct. 22, 1979 revealed that Chrysler Corporation Cars, Mercedes, Cadillac, BMW and the Datsun models tested were amongst the highest nitrosamine emitters with Rolls Royce and Bentley being amongst the lowest.

In the January 1976 issue of SCIENCE, an article discussing nitrosamines concluded that even though there was no direct link to their producing cancer in humans, the concern that they could possibly produce cancer did not stand out as clearly and unequivocally as did those studies on vinyl chloride, philosophizing however, "that it is reasonable to assume that man is not a god and that if every animal group tested has succumbed to it (nitrosamine-induced cancer) then man will succumb as well."

A lengthy and introspective article on the new car smell linked phthalates used in plasticizers to experimental and human health problems.[12] These compounds are related to thalidomide which subsequently was proven to cause serious birth defects in humans and withdrawn from the pharmaceutical market and only recently re-introduced as an anti-cancer drug. It was found that blood stored in polyvinyl chloride bags released phthalates after being stored for the usual 21 day storage period and were possibly implicated in the production of shock-lung in patients receiving large amounts of transfused blood. One possible mechanism being the clumping of blood platelets which in turn would produce small clots within various organs of the body. Other chemicals used in plastics other than plasticizers are stabilizers and dyes, each having their own inherent set of problem-producing complications in humans. Researchers pondered the question of degradation and interaction of combinations of chemicals in the various types of plastics as possibly being more dangerous than any element alone.

Various common stabilizers used in the manufacture of plastics are known chemical toxins. The cloudy film deposited on the interior windows of a new car is due to plasticizers, and is common in hot weather, as the plastics exhale their toxic fumes into the occupants' lungs and throughout the car's interior. As a matter of fact NASA (National Aeronautics and Space Administration) has banned the use of polyvinyl chloride plastics in space capsules and research vehicles because the plasticizers condensed on extremely sensitive optical equipment.

In the early 1980's, a plaintiff won the case and a new car when he filed suit against the Ford motor company because of plasticizers forming on the car's windshield.

In December of 1983, Nissan responded to the Attorney General of North Carolina regarding a customer's complaint of excessive interior odor and film forming on the interior windows of her Datsun. The letter replied that the problem was common and harmless, as corroborated by the Federal Food and Drug Administration, but they would kindly replace the vinyl on the seats and door panels with new original equipment parts further stating however, that "their offer is not to be construed as an admission of liability on the part of Nissan."

Chevrolet sent a dealer bulletin to its retailers in May 1972 stating that "the migration of materials used in the manufacture of plastics on interior windows is to be considered a normal maintenance function, and can be removed with glass cleaners containing ammonia or vinegar."

The EPA in September 1987 issued a technical bulletin noting various toxins contained in the air samples of various vehicles. They identified 147 compounds among the carcinogenic compounds identified as follows:

> nitrosamines
>
> aniline
>
> dibromomethane
>
> dichlorobenzene

dimethyl phenol

isobutyl alcohol

maleic anhydride

benzene

carbon tetrachloride

chloroform

phenol

The report concluded that since the exposure level was felt to be low it was thought that no significant risk could be occurring as a result of exposure to these substances. Sounds familiar, doesn't it? Obviously, our whole environment is generally very safe and regulations are being strictly enforced and obeyed by industry and the corporate giants. That's also why we can find snowballs in Hell! If all of this were just so, then why is our population dying by leaps and bounds? We are succumbing to a multitude of cancers and increasing cardiovascular and pulmonary diseases at alarming rates, all of which are increasing annually. The last sentence in that EPA report states: "No further work is planned to identify other substances in vehicle interiors at this time."

After having a few letters to editors concerning the new car smell published in car magazines in the United States and England, I received a number of letters which I will now relate to the concerned consumers and victims of the automobile industry and its suppliers.

The editor of Car and Driver sent me a letter from one of its readers who shared interest in my report on the new car smell in my Acura Legend. This young man was a prominent business executive who owned a 1989 BMW convertible which was 2 years old. He stated that he and his fiancée suffered from headaches, dizziness, weakness and pains in the head other than headaches, just as they did when entering certain buildings.

He had the air analyzed and it was found to contain high amounts of methylene chloride, the concentration of which was not quantified at the time of his letter. It was believed to be coming from the ducts of the vehicle. Sounds familiar, but please wait until the next section. My wife and I, as previously stated, became ill with similar symptoms while briefly driving a BMW 318i loaner with plastic seating while our 850 was being attended to.

I was only able to tell this person that he was correct, many people have complained of similar symptoms and sickness from the new car smell, and that I was contacting many auto manufacturers concerning this problem, which was more severe in cars having plastic seating.

A personable young woman, 36 years old, was referred to me by a Toyota Dealership in New Jersey after she became ill in two cars, an Acura Integra and a Toyota Corolla. Shortly after driving the cars from the dealership, she became sick within minutes, complaining of tingling of the lips and mouth, dizziness, light-headedness, then shortness of breath if she remained in the car. In 1991, I received one of many letters concerning this problem.

A man sent a letter requesting a phone conversation concerning his very expensive, top of the line, Lexus SC 400 coupe which was only 3 months old, with 2,000 miles on the odometer. He was constantly lightheaded and dizzy since he purchased the Lexus coupe. I discussed the problem with him, told him that my wife and I had similar problems after testing all the Lexus models, which was the only reason we never bought one.

One of the more interesting stories was that of a businessman who purchased his wife an Acura Legend. I believe it was a Christmas present, but I've since forgotten. The story came from the leading salesperson of the Acura line in that particular dealership. The woman became intensely ill every time she drove the car. It took her a few weeks to decide that it was the car. She and the salesman decided it was due mainly to fumes coming from the ventilation ducts. She was given another Acura, and the same thing was happening. She finally gave the car back for another brand. I tried to contact the husband who purchased the car for his wife, but he was in jail for some infraction involving a business deal or something of the sort. She did assure us that her husband was not trying to get rid of her – story coming up. Neither was I trying to get rid of my wife when I purchased her an Acura Legend coupe for her birthday. What a near tragedy that turned out to be!

The Commonwealth Scientific and Research Organization (Australia) published a report on the new car smell in December of 2001 which was highlighted in the Wall Street Journal on April 18, 2002. The report concluded that toxic emissions in new motor vehicles are found for six months and longer after the initial purchase and that the levels of these toxic fumes were many more times the goals established by Australia's National Health and Medical Research Council.

In controlled exposures on human subjects, they found that concentration of these toxins at levels only one-half that found in new cars caused symptoms within minutes. These included various feelings of discomfort, drowsiness, fatigue, confusion, eye, nose and throat irritation, headaches, and last but not least, neuro-behavioral impairment. Story coming up. In addition to those previously mentioned were included: acetone, cyclohexanone, N-hexane, styrene, toluene and xylene. The researcher aptly and appropriately called it a "toxic cocktail." Shockingly, but not very surprising, was the conclusion that air pollution costs the Australian community in excess of 10 billion dollars a year in lost productivity and illness!

The Scientific Instrument Services Corporation (SIS) studied air samples in a 1995 Lincoln Continental, and over 100 volatile organic compounds in the air samples analyzed, and confirmed previous studies that although the levels decreased over time, increasing ambient and interior car temperatures caused the levels of these compounds to rise. There was no mention of long-term testing to see just how long these emissions would continue to plague the occupants.

The public is finally being made aware of the seriousness of that toxic new car smell. The manufacturers are now keenly aware of the harmful probabilities of the noxious new car smell. Some had begun several years ago to remedy or ease the situation to one of a less noxious degree, still others are reluctant to change the deplorable situation in their cars. Buyer beware. And if you are aware, and stay away from those types of vehicles, your vote will be counted, sooner or later.

Basil M. RuDusky M.D.

Part V

THE ACURA LEGEND STORY

Basil M. RuDusky M.D.

THE ACURA LEGEND STORY

It was early spring of 1991, a year which was to be one of the hottest in the weather annals of the Northeast, and Mountaintop, Pennsylvania was not to be exempt from the soon to arrive summer heat. It would be a summer which my wife and I will never forget. I purchased her a luxury car which she could call her own. She was content in driving one of my second or third cars, but I felt that she would enjoy having a car she could call her own. We were impressed by the style and quality of the new Acura Legend coupe. We purchased one painted a stylish looking metallic quartz gray with matching leather. Coming home from the dealer, we did notice and comment on the excessive new-car smell, but felt certain that it would dissipate over a brief period of time, especially with the arrival of the hot weather, using the air-conditioning system most of the time, and keeping the windows down as much as possible. Little did we realize just how wrong we would be! A nightmare was about to unfold. After all, up to this point in time we owned about 3 dozen cars, and none of them ever made us ill, although admittedly, none ever smelled as badly as this one. We certainly did not think what we purchased would affect our health and lives to the serious extent which it had. Most certainly, we never expected these effects to be permanent and everlasting, as they have been to this very day. My wife is a registered nurse – observant, non-complaining and was always healthy. Even when ill, she would never complain of anything. She usually kept everything to herself. After her driving the car for approximately 2 weeks, she reluctantly complained that she was not feeling well. Her symptoms were certainly serious enough and I knew, shouldn't be ignored. She complained of weakness, fatigue, burning in the chest, some loss of balance, headache and even worse, memory loss and difficulty with concentration. She is not one for going to doctors, but generally relied upon me for medical assistance, as do

I myself. She felt as though she may have to be hospitalized for testing and evaluation. Admittedly, I was truly frightened, for she had never complained of any significant health problems in the past, nor was she so eager to seek help. I feared the worst. The symptoms were compatible with a central nervous system problem, many of which can be permanent, crippling or fatal. After two days of discussion and deep contemplation, she replied – "maybe it's the car." I said that would be easy enough to solve – don't drive the car, use mine, and I'll drive your car, then we can switch again in order to attempt reproducing the symptoms in one of us. Sure enough, after 2 to 3 days, I began to get similar symptoms, just as bad, if not worse. They consisted of weakness, lightheadedness, dizziness, nausea, lethargy, sleep disturbance, headaches, burning in the chest and throat, cough, blurred vision and memory loss, accompanied by difficulty in concentration. The difficulty in concentration was particularly difficult to assess because I felt so sick and generally upset in an almost unexplainable way that you couldn't concentrate on anything other than feeling so badly.

The problem was solved, and there was no point in giving the car back to my wife, and I certainly could not drive it. It took 3 weeks for her symptoms and feelings to clear before she became her usual normal self again. This all occurred within the first 2 weeks of ownership. I called and wrote several letters to the Acura division of Honda Motors, based in Torrance, California, discussing the problem as well as the erratic idling of the car and its sharp, fast jerk when put into gear with the automatic transmission, indicating that the latter was dangerous, and the former a health hazard. Being stubborn, persistent and as called by a research colleague who described me as being one of the most tenacious clinicians he has ever encountered, I felt that there was no problem that could not be adequately remedied with time and effort. So the process and rituals began.

The first week of May, 1991, I placed the car in a huge storage garage of 7200 square feet, all windows, sunroof and doors open, with the battery disconnected.

After 8 days, I entered the high-ceiling metal structure with a concrete floor, and upon entering, it became immediately apparent that the stench from the car interior completely filled the entire garage – mind you, a 60 x 120 foot structure with an 18-foot ceiling! Worse than I thought, I pondered. I sat in the car a few minutes, became lightheaded, slightly dizzy, had irritation of the chest with cough, and that night had difficulty sleeping. I was really ill once again. My wife was correct. I began placing calls to various consumer and environmental protection agencies, as well as to my congressman. I became more distressed after knowing of others who complained about similar problems with this highly-touted and much over-rated car line.

The third week of May, I decided to give it another try after putting it away to detoxify itself. Within 1 to 2 hours, the symptoms again returned, this time with a chemical taste in my mouth, similar to paint fumes, plus most of the above-mentioned symptoms.

By this time, I realized that the car was big trouble, a real hazard to one's health. But, being persistent, and unwilling to give up the ship (Acura's flagship), I decided to capitalize on the "long, hot summer" of 1991.

I will not present my day-to-day and week-to-week notations, according to date and time of day or night as revealed in my Acura diary, but will highlight and summarize my actions and reactions to this painstaking endeavor which took place during that hottest of summers. Obviously, storage of the car in a huge garage was offering no benefit, so I wisely surmised that the car needed heat and sun. Acura certainly was not going to do anything, they could admit no fault, nor could their representative appreciate the blatant seriousness of the problem. The federal government was idle as usual, when it comes to inexpensive and appropriate legislation for the benefit of the working class, and I had no time to waste on dealing with lawyers, so I was left with the most practical solution. I would make or break the car.

On a daily basis, rain days (few and far between that summer) excluded, the car would sit in the sun and air out all day and evening, sometimes from 8 am to 11 pm, windows down, sun-roof open and eventually, trunk and hood open as well. On returning to the garage late at night, the weather still being intensely hot and humid, we could smell toxic fumes coming from the car 10 to 15 yards away!

Finally, I decided, this is it, all out war. It was me and everything I could conceive of against this purveyor of excessively noxious chemical fumes and vapors. I still believed I could remedy the problem and thereby hopefully keep and enjoy driving this attractive coupe. I began using a mixture of a neutral car wash soap and water, and on a daily basis, would spray the entire car interior – and I mean everything. The car would actually be drenched – I then proceeded to wipe the entire interior, and allow the car to remain outdoors the rest of the evening and night – even overnight. Actually, I really didn't care if anyone would steal the car, for I'd be glad to get rid of it. I even thought of setting it on fire! But I figured, easier to negotiate a trade than deal with the insurance company. Frankly, I'm quite surprised the car didn't ignite itself, for the next part of the battle was really extreme. In between all of these superhuman efforts, of which I was certain no other person in the entire world would ever undertake, I would drive the car to work, and invariably, in spite of the decreasing, but immediately detectable fumes, I would become ill within 10 minutes or so – dryness of the nose, mouth and throat, burning in the chest, lightheadedness, cough, an unexplainable sick feeling accompanied by a strange feeling in my brain – all with fresh air coming into the car with one or more windows down! I also discovered an additional source of the toxic problem.

What I found was that something from the initial startup of the car was causing the immediate reaction, in spite of the fact that the nose-detectable interior car smell was diminishing. I then decided to place my nose against the various dash vents after immediately starting the car. I was hit in the face with a blast of noxious fumes coming from the vents! I could only describe it as a pungent, acid-like odor, highly and immediately irritating, which smelled something like formic acid, which was the only way I could describe it. I could not exclude the frightening possibility of exhaust fumes leaking into the ventilation system nor separate that possibility from toxic fumes emanating from the material from which the ducts were made. Another letter off to Acura and the usual negative response, with many unanswered phone calls. We were now in mid-July of 1991. On one of their rare replies by phone and letter, I was told to have the entire duct system torn apart "at my expense, and prove that there was a problem", then they'll correct it. Sure guys, a few more thousand dollars wasted and I'll end up the same way – a traded, new, as off-showroom floor car, but now with even greater loss on a trade-in.

We were now in the hottest month of the summer – August. It was scorching. Still continuing to be disgusted and stubbornly tenacious, I decided upon the fourth and final step in the battle against this terribly noxious new car smell.

It was to be the battle against all odds. The Waterloo of Cardom. This was it, I thought to myself as I conceived the plan for the final battle. Still in the blazing hot summer sun, I did all of that which I had been doing on a daily basis, only now I added the cavalry to the battle, hoping it would turn the tide toward victory.

Each morning I would start the car, turn the heating system to its maximum (90F) and on alternate days would change the airflow between the interior air ducts to the defogger ducts. I must remind you that the car would idle for 5 to 8 hours a day, fan blower on high, heating system on high, with no one checking the car in the meantime. When I would return to the car, the entire car was as hot as hell. Why it never melted the interior or caught on fire, is itself a miracle. You could fry eggs on that fume-emitting vinyl dash. Opening the doors would burn your hands, and only twice did I find that the engine had stalled. It kept running and purring like a Swiss watch. On running it on the highways it never faltered, never overheated and never lost power. Still, I kept getting ill after being in the car a half hour or less. The symptoms within my chest, head and entire body would last for 1 to 2 days after staying out of the car. This went on through the continued hot months of September and October.

I knew then, that the battle was lost. I had met my Waterloo, and that no such battle had ever been fought, nor will it ever happen again. Finally, the end of October, the monster was traded in on the newly arrived 1992 Cadillac Eldorado, which although somewhat stinky, did not make my wife nor I ill in any way. At a cost of much personal time, effort and labor, plus an $8,800 dollar loss on the trade, I learned a valuable lesson about the new car smell, especially cars that come from the Far East. Unfortunately, as with many other people who have become ill from occupational and environmental exposure, we had become sensitized to some of those noxious fumes emitted by the Acura, and within minutes, begin to get symptoms when sitting in various new cars. At least now, because of that unfortunate and inexcusable experience we now know which cars we can buy and tolerate. We have since purchased and have driven 14 cars with no difficulty – all German and American luxury cars, including the most recent Volvo S-80 turbo, Cadillac STS and Audi A-6 – all with no symptoms, unless we sat in cars which were obviously harmful to one's health, mainly some of the American cars as well as the cars from Japan and the other Asian manufacturers.

Price category made little difference. The cars could be of average cost or very high cost, and the new car smell would be there, sickening and to us, frightening, for we now felt sorry for those unaware of the possible harm being done to their bodies and bodily functions.

Sadly, in addition, we have become sensitized to various chemical inhalants, thanks to Honda –Acura. I can no longer use certain car-care products. We begin to feel ill when exposed to certain household items, furniture, and carpets and when we walk through various buildings that contain new materials and furnishings. When I am exposed to automobile and diesel fumes, especially the latter, I become ill for 2 to 3 days. None of this had ever occurred prior to the sensitization caused by the Acura experience. Neither of us were cigarette smokers, nor did we have any allergic or pulmonary problems, and we were middle-aged when this occurred. We have become the best of automatic toxic air pollution detectors. We can immediately sense noxious chemical fumes in the air no matter where we are, just as we have become automatic good versus bad seat detectors which we can often do within minutes, or in the borderline cases, after driving the car for a few hours or few days at the very most. All learned from personal experience, at a very high cost, accompanied by some sad experiences and some that were pleasurable, if not fun. I never realized that learning could be so costly, but it was also educating and interesting. We can only hope that more serious medical harm will not befall us as time passes by. We are constantly fearful of developing any of those numerous maladies which have been proposed and suggested by the research which has been done to date. Thus far, I now suffer from chronic bronchial sensitivity syndrome, which, although most annoying and unpleasant, to the best of my knowledge is not fatal. It can however, lead to the development of bronchial asthma, and presumably predisposes one more easily to the contraction of pneumonia. Both of these conditions can be life-threatening of course, but not nearly as much as could the development of lung cancer and other possibilities, including changes in one's central nervous system (brain, spinal canal and nerves). Perhaps now, for the mere cost of an inexpensive book, these lessons can be learned by many, since the way was paved by much money and grief by two ardent car-lovers.

Part VI

FINAL THOUGHTS AND ADVICE

Basil M. RuDusky M.D.

FINAL THOUGHTS AND ADVICE

In conclusion, there is no doubt that the car you buy is of great importance to the short and long-term status of your health and subsequently, your wealth, for both are involved in one's well-being. The health of your back, your lungs, and your body are dependent to a significant and generally unapparent degree, upon the car you drive, and is an important determining factor in the continuance of your health.

Knowledge and awareness of what it takes to keep that health and ultimate wealth as protected and functioning as well as possible under the present circumstances of our modern environment and its social structure can lead to definite health and economic benefits to the consumer.

When buying any car, new or used, or any vehicle you intend to drive for that matter, awaken and pay attention to your God-given senses. The senses of sight, smell and feel. Start with the car seat, after a certain amount of experience, you can weed out the bad from the possibly good to good seats by sight alone. Check for proper positioning and the comfort quality of the headrest. If that is inadequate, you saved yourself much time and trouble – forget the car, and go to another make or model. If the upper portion of the seatback has no forward tilt, you must pay very strict attention to the seat adjustments and positioning, for you will need a lot of adjustments to get proper back comfort, and this can be had only with the highly advanced technical system offered by the multi-variable electrical adjustments needed for both, driver and passenger. If your income bracket limits you to a car without these technical advantages, then you must be much more careful to assess the proper design and construction characteristics of your seat. It can be done with the very inexpensive manual adjustments if you have the basic foundations of a good seat. Those foundations are shape, springing, padding, size and design and stitching of the covering material of the seat. I cited the 1993 Infiniti J-30 as one of the world's best automobile seats, in spite of the fact that it had very few adjustments, and no lumbar support, yet it was like "being in heaven" to sit in. If your seat has too much "bolstering", like the magazine car-tester nuts desire, especially if their construction is very stiff, and often times very hard, your body will pay the price every time you enter and exit your vehicle, mainly on exit. You will strain your back, injure your muscles, traumatize your sciatic nerve, and if a male, can also injure your "jewels" as well – one or both. And remember, if you are not young, or already have a back problem, a leather seat is a necessity.

Once you've decided that the car in which you are interested has a seat which is acceptable, you must look at the controls on the dash.

Look at the ventilation controls and check if they include that very important semicircular arrow, indicating that the car has a recirculation air system to prevent outside air from entering the vehicle. (Figures one and two). Next, check with the sales representative and look in the brochure of the model which you are interested in purchasing, in order to determine whether or not the car has a ventilation filter. Most books will state that a microfilter, or electrostatic filtration system is standard or optional on the car. Remember, the simpler systems that state microfilter, dust and pollen filter or particle filter, do just that – they only filter out dust, dirt, and pollen. They do nothing to filter out the many noxious fumes to which you will be exposed. To have some of these rewards, you need a system that states it has an activated charcoal, charcoal, or activated carbon filtration system or something similar. If the car has neither, the best advice is to forget it, and look for one that at least gives your health some measure of protection.

While checking out the seat as it applies to your particular body build and habitus, and your checking on the ventilation system, pay close attention to your sense of smell, and your brain. If the car interior smells excessively, stay in it for as long as you can, doors shut, windows up and pay attention to your nose, your mouth (tongue and saliva), and eventually your brain. That is the order in which you are affected and the order in which they register in your brain. Once you can taste it, you are in real trouble. Once your brain begins to feel it, you will be in big-time trouble, which can be very serious to your overall health and well-being.

All of this can be done without taking the car for a test-drive. If you are satisfied up to this point, or perhaps doubtful, take the car for a long test drive, for at least 30 to 60 minutes of driving as well as sitting on the passenger side. Include all types of roads, the bumpier the better, in order to feel the amount of shock transmitted to your back from the suspension through the seat. Check to see if the low speed steering is easy enough for you to avoid straining your neck, shoulders, elbows and wrists. Pay attention to the signals coming from your back. If you feel shock, strain, or get out of the car feeling stiff and achy, forget it.

Then drive behind the foulest, stinkiest vehicles you can find, which in order are: school buses, regular buses, diesel trucks, gasoline-engine SUV's, vans and older beat-up cars. Do so first with the recirculation air system on, in order to determine its effectiveness, then, if it has an activated charcoal ventilation filtration system, with the recirc button off, so that you can determine the effectiveness of the fume extraction system. It is irrelevant to do so with a microfilter, as it will afford very little fume extraction and protection.

Throughout the test-drive, and when returning, stay in the car for as long as you can in order to determine what effects the new car smell has on your body. Every person is different. Each will respond differently and over a different period of time. Some individuals even think they are in paradise, not realizing that they are being slowly poisoned by dozens upon dozens of toxic chemical fumes. Pay strict attention to your nose, eyes, mouth, saliva, chest and brain. A signal from any one of these sources means eventual trouble for you – forget the car, and start the process all over again. Sooner or later, you will become an efficient, automatic, car-testing, car-tasting and car-smelling machine, and be the better for it in the long run of life, so why shorten it by being inattentive. Buona fortuna and buon viaggio, as they say in Italian car-buying circles.

Part VII

ADDITIONAL HELPFUL HINTS

Basil M. RuDusky M.D.

ADDITIONAL HELPFUL HINTS

The basic tenets of car buying must include the buyers' ability and perseverance in seeking out the necessary information and asking the appropriate questions. It is advisable to get the latest brochure on the automobiles being considered. Read it thoroughly and then discuss the particular information sought and questions to be asked with a knowledgeable salesperson (a rare species), the service manager (an equally rare species) or the manufacturer representatives (an even rarer species).

It is sometimes helpful to ask to see the owner's manual for the particular car being considered, as is sometimes the case, it will tell you the type of ventilation filter which is present in that model. Unfortunately, only the engineers in the ventilation department know the effectiveness of the filter being utilized, as well as the type of filter (cloth, foam, microfilter, electrostatic filter, charcoal, charcoal-impregnated, etc.). They have information from the manufacturers of the filters that state the particular particle size each filter is capable of removing from the incoming air. Any filter that allows passage of particles 5 microns and larger, are basically worthless. The better microfilters are capable of filtering particles of 3 microns or less, so that they can also remove some airborne bacteria, in addition to the usual dust, spores, fungal particles and pollens.

As previously stated, any filter is better than none at all, and must be accompanied by a true air recirculation system. It must be remembered that only an activated charcoal system or its equivalent is able to remove some of the harmful odors and fumes which so commonly plague our nation and the world in general.

Depending upon the construction and location of the filter, a little ingenuity can go a long way in allowing you to improve the standard equipment filters, not only with regard to particulate matter, but fumes as well. When buying a used car, check the information on the particular make and model carefully in order to determine if it was equipped with a ventilation filter. Shortly after purchasing the car, have the ventilation filter changed, requesting the dealer to give you the filter being replaced. If it is new, box it, and save it for the next year. Receiving a used car with a newly replaced ventilation filter would be a miracle.

A call to the automobile manufacturer can sometimes be helpful. The toll-free phone numbers are usually listed in the brochure. I must warn you in advance however, it will often be a painful experience. The individuals at the other end of the line are usually, and generally not only unknowledgeable about automobiles in general, but also know little or nothing about the products they are supposed to represent. Some even give the impression that they do not want to work and resent your taking up their time. Most of them only have a brochure in front of them, and some have an inadequate specification book to go by, which when combined with their lack of knowledge and desire, leave the consumer in a state of disappointment, it not resentful disgust with the person as well as the system. You soon begin to realize that they are being paid good salaries for taking up space and little else, except to placate the inquisitive consumer.

Occasionally, you will come across a person that is truly interested in doing his or her job, and will try to help. These individuals will contact a technical services person and seek the answers to your questions. The last resort is to write a letter to the appropriate division of the auto manufacturer. You may get information as to the effectiveness of the ventilation filter but usually the response will be ignored.

On the other end of the medical spectrum, the new car smell, it is unknown to the consumer or consumer advocate as to whether or not a particular automaker is testing the interior smell of their products at all, or if so, just how efficiently it is being done. Usually you will get no answer or a political-like form letter stating they are interested in the problem and are addressing same, with no scientific information being given to anyone, probably not even to the federal government, who in turn, could be less concerned than the automakers themselves. So you are left on your own to be the judge of what you should buy or refrain from buying. With the renewed world interest in the possible and probable hazards of the new car smell, I am sure that most concerned and conscientious manufacturers are turning some of their attention to the matter.

While it is true that many individuals are tolerant of the new car smell and seemingly unaffected by it, there are important considerations to be dealt with.

The first and foremost problem is that which involves persons who are already sensitized to any or most of the volatile fumes involved. These individuals usually already suffer from the so-called "sick-building" syndrome, and will react immediately and severely to the new car smell. The second problem of concern, but equally as important as the above, involves those individuals who have various disease states which can be and have been proven to be, aggravated by the new-car smell. These are the patients with the various lung and heart problems previously discussed. The third problem is an unknown factor – just how much harm is being done to a healthy body? The answer of course is unknown. Finally, there will be a significant number of persons who, unfortunately and unknown to themselves, will become permanently sensitized by overexposure to an excessive amount of new car smell. Note the word "excessive", an undefined quantity, which when known even by the manufacturer, goes unreported. As in the near tragic cases described it only takes several relatively unprolonged exposures to result in permanent sensitivity to the substances involved. Once present, it will be a lifetime affair, and often becomes worse rather than improve. Therefore, if you open the car door and the fumes hit you in the face, or you sit in the car a few minutes and begin getting a "high", or feeling strange, you better get out and keep looking. It is the only way you can avoid being affected or injured by one of the aforementioned problems.

If your judgment after buying the wrong car was conclusive to your satisfaction that the noxious fumes were causing you to become ill, you can request that the dealer exchange the car for another make (if he handles more than one line of automobiles), or ask the manufacturer's representative to take the car back and refund full payment (a rare concession). Once you are sure the car is making you ill, you must document by date, symptoms and time, each repetitive occurrence. Witnessed documentation by another individual is also advisable. Remember, what can be noxiously smelled, can be retrieved and tested. If it is your decision to go "all out", rather than lose considerable money by trading in the car and buying another make, you will then need to proceed with a highly reputable industrial pollution and toxicology testing laboratory to analyze the fumes emitted from the various elements used in automobile production which cause the new-car smell. This can be costly, but once you've decided to go that far, you will likely need the assistance of a reasonably competent and preferably experienced product attorney. You can utilize the data in part three as a guide to the chemicals generally considered in an analysis performed under these circumstances. You can be assured that when properly done, anything that you and someone else can instantly smell, and simultaneously induces you to feel ill, can be detected in quantities that will far exceed all industrially-and governmentally-regulated standards for personal and environmental health and safety. And remember, the last thing any automaker wants in these days of keen competition is to foster a bad reputation amongst the car-buying public. It is just bad business, and they are finally beginning to realize that simple fact of life in our present society. The consumers in general have become more interested, and more concerned about what they buy, and more knowledgeable about what they are buying. Good luck and good hunting – may the force be with you.

Appendix

Cars Evaluated in 2002*

CODE:

Seats and Headrests**	Interior Smell	Ventilation Filter
E = excellent	N=negligible or none	C=charcoal / carbon
VG = very good	S = slight	M = microfilter
G = good	M = moderate	O = none
F = fair	B = bad	? = unknown
P = poor		

	Seats	Headrests	Smell	Filter
BMW 5 series	E	E	N	C
BMW 7 series	E	E	N	C
BMW 3 series	E	E	N	C
BMW Z-3	G	G	N	M
BMW Z-8	P	P	N	M
BMW X-5	G (1)	E	N	C
Audi A-6	VG	G	S	C
Audi A-4	VG	G	S	C
Audi A-8	E	E	N	C
Audi TT	G	G	N	M
Porsche 911	G	F	N	C
Porsche Boxster	F	P	N	C
Cadillac Deville DTS	VG	G	S	M
Cadillac CTS	F	P	M	M

Cadillac STS	VG	P	S	C
Mercedes S Class	E	E	N	C
Mercedes E Class	VG (1)	G	S	C
Mercedes CLK	G (1)	F	S	C
Mercedes SLK	G	P	N	M
Volvo S-80	VG (3)	G (5)	S	C (6)
Volvo S-80 T-6	VG (3)	G (5)	S	C (6)
Volvo S-40	F (3)	P (5)	S	M
Volvo S-60	VG (3)	VG (5)	S	M
Volvo S-70	E (3)	VG (5)	S	M
Lexus ES /300	G	P	B	M
Lexus RX 300	G	G	B	M
Lexus IS 300	G	P	B	M
Lexus GS 300	G	G	B	M
Lexus LS 430	VG	VG	B	C
Toyota Camry SLE	G	G	B	M
Chrysler 300M	VG (3)	G	N	O
Jeep Grand Cherokee	G	F	S	O
Hyundai Sdn	(4)	—	B	O
Hyundai Tiburon	(4)	—	B	O
Hyundai X350L	G	F	B	M
Saturn L3000	VG	G	S	M
Acura 320L	(4)	—	B	M
GMC Envoy	G	P	M	C
Acura RL	P	P	B	M
Acura RSX	(4)	—	B	M

Honda Accord	(4)	—	B	M
Infiniti I-30	E	E	S	M
Infiniti Q45	VG	VG	S	M
Jaguar X series	F	F	S	M
Jaguar XK	VG	E	S	M
Corvette	VG	VG	M	O
Saab 9-5 Aero	E	E	N	C
VW Beetle	VG	VG	S	C
VW Passat	VG	VG	S	C
VW Jetta	VG	VG	S	C
Buick Rendezvous	P	P	M	C
Subaru Outback	F	P	S	M
Ford Taurus	VG	P (2,7)	S	M
Mercury Sable	VG	P (2,7)	M	M
Mercury Cougar	G	P (2,7)	M	O
Lincoln LS	G	P (2,7)	S	M
Lincoln Town Car	G	P (2,7)	S	O
Lincoln Continental	G	P (2,7)	S	M

* Based on personal subjective evaluation of 2 individuals.
* * Evaluation for comfort only.

1. hard side bolsters
2. too far posterior
3. bottom too short
4. interior smell so bad, did not enter vehicle
5. headrest too posterior in spite of anterior tilt, due to lack of anterior tilt of upper portion of seat
6. beginning with 2002 models, previously had microfilter
7. no fore-aft movement of headrest

Basil M. RuDusky M.D.

BEST SEATS EVALUATED IN 2002

<u>World's Best Seat:</u> BMW 7 series*

1. BMW

2. Mercedes

3. Saab

4. Volvo

5. Audi

6. VW

7. Infiniti

8. Cadillac

* based on author's personal judgment & experience.

"CLEANEST" INTERIOR NEW CAR SMELL, 2002 *

1. BMW
2. Mercedes
3. Saab
4. Volvo
5. Audi
6. VW

* based on objective evaluation of two persons

BEST INTERIOR AIR QUALITY, 2002 *

1. BMW
2. Mercedes
3. Audi
4. Saab
5. VW
6. Volvo

* based on evaluation and road-testing, manufacturers data and opinion of author

NEEDED FEDERALLY-MANDATED
REGULATIONS

1. Specifically restricted anti-idling laws for all motor vehicles, in increasing degrees, dependent upon type of vehicle and engines.

2. Anti-pollution regulations on all construction and service equipment.

3. Anti-pollution regulations for all small engine vehicles and appliances, including motorcycles, lawn mowers, snow-blowers, weed-cutters, etc.

4. Stricter regulations regarding the formulation of gasoline and diesel fuels.

5. Nationwide smoking ban in all public places: workplace, restaurants, bars, streets, parks, entrances and exits of buildings etc. Smoking to be

allowed only in ones' personal vehicle, home or yard.

6. Elimination of insurance benefits, disability, workmens' compensation, etc. for all self-inflicted diseases due to cigarette smoking, alcohol and drugs.

7. Increased regulations regarding environmental pollution by business and industry at all levels, from the smallest business to the largest manufacturers and suppliers.

8. Legislated minimal standards for automotive ventilation filtration systems regarding particle size and fume extraction.

9. Enforced appropriate measures regarding emissions requirements for all vehicles registered in every state; those that cannot meet the standards must be scrapped within a given period of time.

Recirculation Air Button

Cadillac STS

Recirculation Air Buttons, Audi A-6

Top button for continual (manual) air recirculation. Bottom button for intermittent (automatic sensing) recirculation.

Additional Example of Recirculation Air System

Representative Ventilation Filters

MAKE/MODEL	**SIZE**	**COST**	**TYPE**
Audi A-6	8 x 12"	$38.60	charcoal
BMW 740i ('93)	6 ½ x 12 ½"	$27.00	microfilter
BMW 850i ('91)	4 ½ x 11 ½"	$19.50	microfilter
Cadillac STS	8 ½ x 10 ½"	$49.46	charcoal
Volvo	10 x 11"	$ 15.08	microfilter

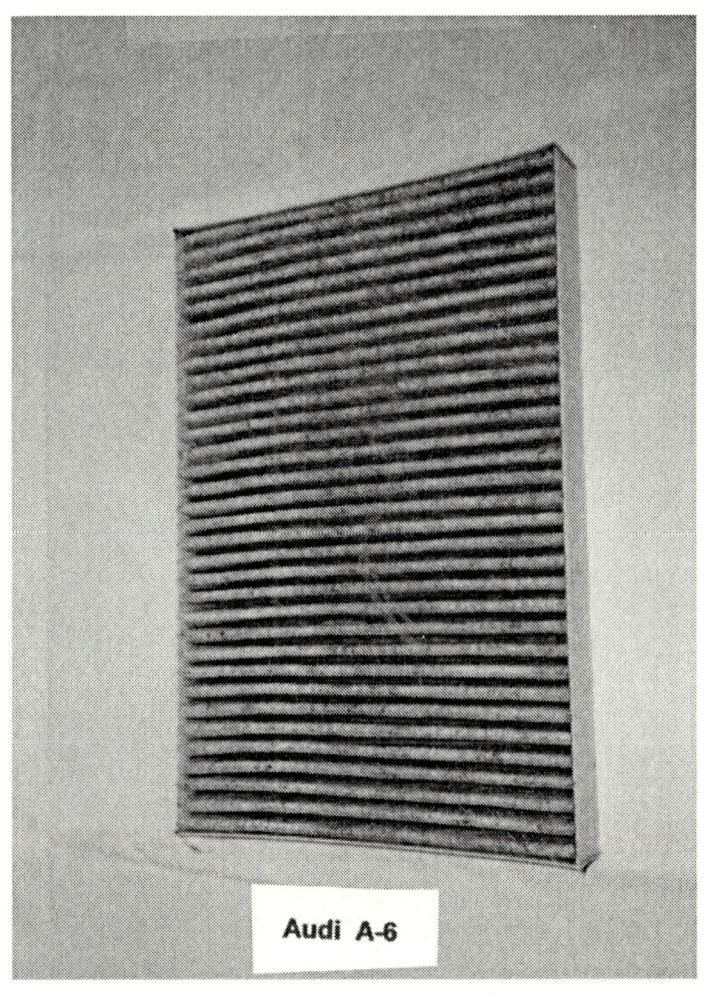

Audi A-6

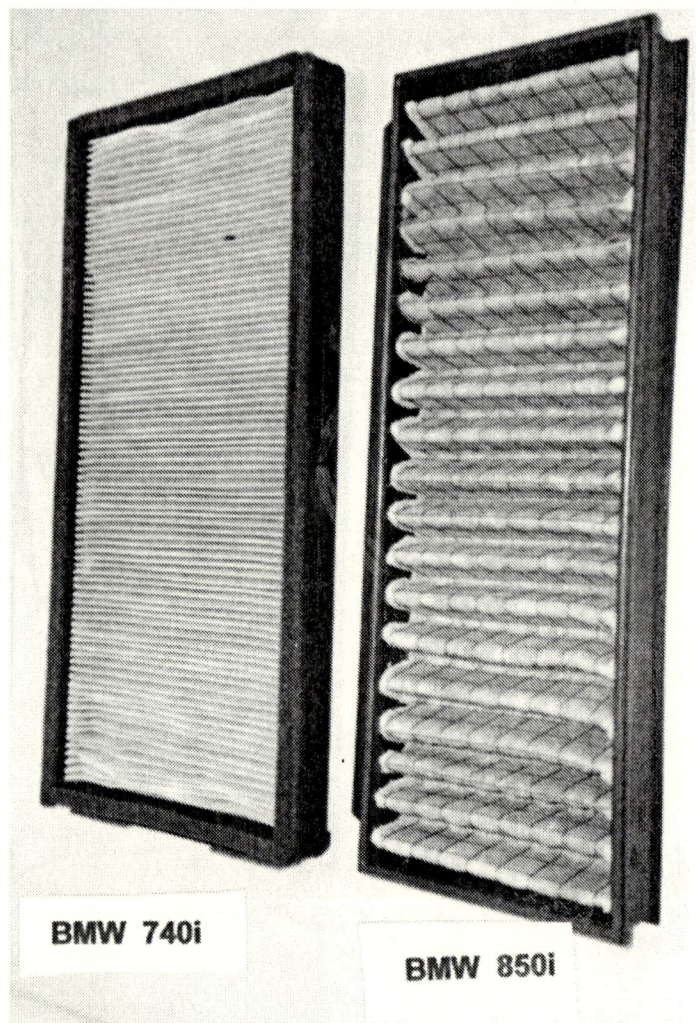

BMW 740i

BMW 850i

Cadillac STS

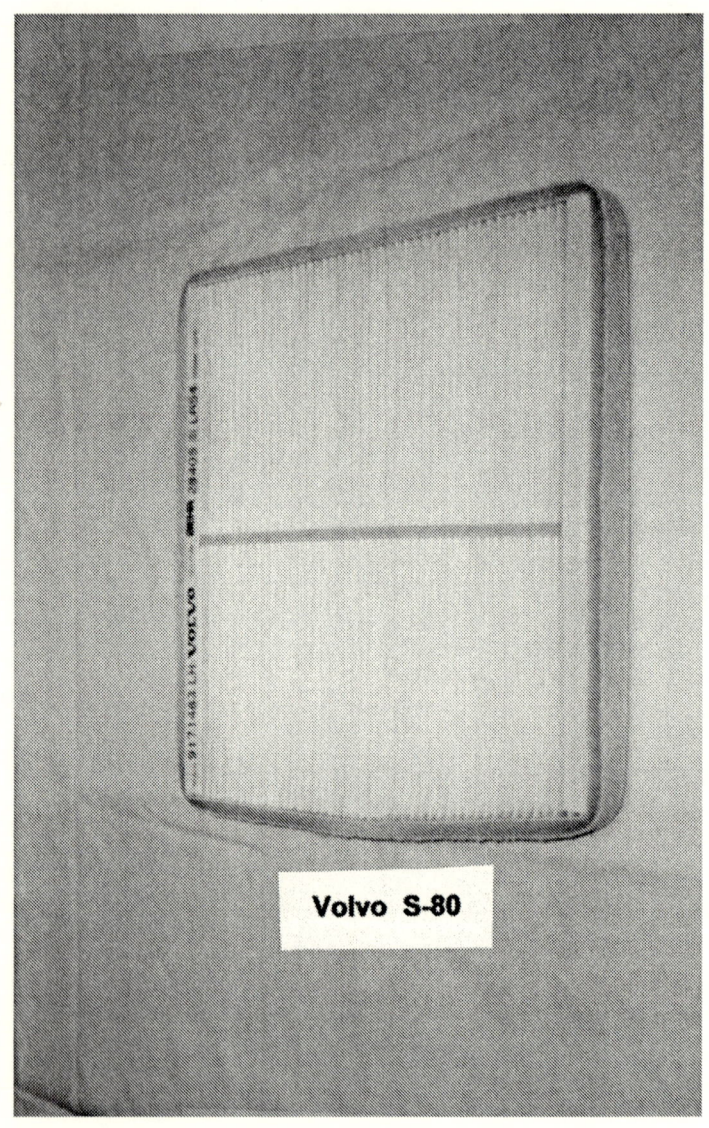

Volvo S-80

References

1. Hashemi L, Webster B, Clancy E, Volinn E: Length of disability and cost of workers compensation low back pain claims. Journal of Occupational Environmental Medicine; 39: 937-945, 1997.

2. Status Report – Insurance Institute for Highway Safety – Whiplash injuries. 30: 1-11, 1995.

3. Klaus, M: Car seats. CBS news, 60 minutes; 24; no.22, Feb. 16, 1992.

4. McMenamin, P: Costs of hayfever in the United States. Ann Allergy 73: 35-39, 1994.

5. Kanner, RE: Urban Air Pollution: Why Is It a Health Problem? CHEST 113: 1161-1162, 1998.

6. Schwartz J, Nees LM: Fine particles are more strongly associated than coarse particles with acute respiratory health effects in schoolchildren. Epidemiology 11: 6-10, 2000.

7. Peters A, et al: Air pollution and incidence of cardiac arrhythmia. Epidemiology 11: 11-17, 2000.

8. Alexander GJ, Kanner RE: Clinical effects of air pollution. Contemp Intern Med 8: 9-14, 1996.

9. Suwa T, et al: Particulate air pollution induces progression of atherosclerosis. JACC 39: 935-945, 2002.

10. Pope CA, et al: Lung cancer, cardiopulmonary mortality, and long-term exposure to fine particulate air pollution. JAMA 287: 1132-1141, 2002.

11. Mittleman MA: Triggers of acute cardiac events. American J Med and Sports Mar-Apr: 99-102, 2002.

12. Shea KP: The New Car Smell. Environment 13: 3-9, 1971.

Additional References and Suggested Reading

Steenland K: Lung cancer and diesel exhaust: a review. Am J Ind Med 10: 189, 1986.

Brunekreef B: Air pollution from truck traffic and lung function in children living near motorways. Epidemiology 8 (3): 298-303, 1997.

Schwartz J: What are people dying of on high air pollution days? Enviorn Res 64 (1): 26-35, 1994

.Dockery DW, et al: An association between air pollution and mortality in six U.S. Cities. New Eng J Med 329 (24): 1753-1759, 1993

.Cackette T: Importance of reducing emission from heavy-duty vehicles. California Air Resources Board, Oct: 5, p.15, 1999.

About the Author

The author is a practicing physician specializing in internal medicine, cardiovascular medicine and forensic medicine. He serves as senior consultant to numerous corporations, governmental agencies and insurance firms. He has published over seventy articles including various aspects of clinical medicine and scientific research.

A lifelong automobile enthusiast, he has achieved an international reputation as a free-lance automotive consultant. His special interests have centered around automobile seats, "clean air" ventilation systems and the new car smell. The efforts he has expended over the past two decades have begun to show positive results in improving all three of the aforementioned concerns.

Printed in the United States
958900001B

9 781403 383082